Mathematics: A Humanistic Approach

G. Joseph Wimbish, Jr.

Oklahoma College of Liberal Arts

Wadsworth Publishing Company, Inc.
Belmont, California

ISBN-0-534-00155-6

L. C. Cat. Card No. 70-189793

Printed in the United States of America

1 2 3 4 5 6 7 8 9 10 - - - 76 75 74 73 72

Table of Contents

Table of Contents

Prelude

Mathematics: A Humanistic Approach examines some basic ideas of mathematics and mathematical methodology with the hope of enlightening liberal arts students in the ways of mathematicians. The students who have used the material in this volume in its earlier, developing forms have usually responded favorably to this approach to understanding mathematics.

The book may be used in a variety of ways: both as the main text for the course, and as supplemental material. Ideally the material in this volume should be used as a study guide to the book of readings, which appears under separate cover. The text will often elaborate on a portion of an article, offer a different perspective, or provide a structure for understanding one of the readings. The text is certainly the author's evaluation of how mathematics relates most directly to the liberal arts.

In a users' manual I will outline more completely some possible ways for an instructor to use this material. The instructor should freely express his point of view and permit his students to develop their own mathematical ideas.

I offer my particular thanks to certain undergraduate students who have seen something they liked in my position, and who have assisted by offering germane criticism. These students are: Wesley Shepard, John Graves, Martha Webb, Tim Miller and Mary Laird. One of my fellow faculty members, Dale Maness, the chairman of the Division of Mathematics and Natural Sciences, encouraged me to write this book so that he might adapt it for his own classes. I also wish to thank my editor, Don Dellen, for his patience with my idiosyncratic behavior patterns.

Chapter One

Mathematics Is a Human Process

"The time has come," the Walrus said,
"To talk of many things:
Of shoes—and ships—and sealing wax—
Of cabbages—and kings—
And why the sea is boiling hot—
And whether pigs have wings."

Through the Looking Glass

Lewis Carroll

We present the following pages in the hope that you will find a new experience in mathematics. Although we have attempted to make the experience as pleasant as possible, we shall not try to delude you with inane and asinine statements such as "Math is fun." In the first place, the collection of symbols, m a t h, will never be used since "math" has several meanings. It is a mowing (an obsolete usage); it is a colloquialism for mathematics; or, it is an order of celibate beggar monks. Mathematics is neither a mowing nor a colloquialism! Moreover, even though many mathematicians are beggars, some are monks and one or two even may be celibate by choice, mathematics itself is not characterized by any of these alternatives.

In the second place, while there are many pleasures to be derived from the study of mathematics, many people cannot or do not wish to involve themselves in such a study. Besides, many experiences exist in this world which are just as pleasurable and far more easily obtained. We certainly don't need to tell you what they are.

Myth and Demyth

How often have you heard that you should study mathematics because (1) It can be fun, (2) It is an important tool to be used in other areas, (3) It teaches you to think. Since we have disposed of the first of these items, let us now examine the remaining two.

There seems to be wide agreement, usually among those who have ceased their formal studies, that everyone (else) should study mathematics. What they do not agree upon is what kind of mathematics and how much. However, many people equate mathematics with one of its subdivisions, arithmetic. It is clear that many times what is intended, therefore, is more and more study (?) of arithmetic. The purpose of this study is that a person increases in speed and accuracy and so excels in some clerk-like employment. However, the proliferation of electronic computational devices has lessened the usefulness of this accomplishment. Now understand, we have no argument with the study of arithmetic, as such; we just do not equate it with mathematics. Anyway, if you have not learned arithmetic by the time you read this book, we certainly will not presume to be able to teach you in these few pages! What you have not learned in approximately twelve years of public school cannot be learned in one short college term.

Moving to the implied intimidation of, "It will help you in other areas," the obvious question is "How will it help me in ____________?" (Fill in your favorite.)

It is obvious that the study of some kind of mathematics is valuable in the natural sciences and engineering. On the other hand, why would anyone want to study mathematics for its own sake, or even more germane, why should a student who has none of the above interests be encouraged to study mathematics? In other words, if you are a typical liberal arts, or humanities student who has no interest in becoming a computer programmer or space-age scientist, why then should you take a course in mathematics? Actually, we know of no overwhelmingly good reason why you should. Of course, we feel much the same way about most other courses of study; there are few categorical imperatives for doing most things. Quite frequently, we do what we do for reasons which in the beginning are completely extraneous. For example, we take a required course in mathematics so that we fulfill the requirements for graduation. As someone has said, we are somewhat unwilling participants in compulsory mis-education.

Because in most colleges some mathematics is required of everyone, much that is irrelevant has been foisted off on the unwitting and unwilling student. Consequently, the student does have a legitimate question when he asks, "What's in it for me?" Since this whole book is devoted to answering that question, forgive us if we avoid answering it directly.

Finally, let us dispose of the last myth on the agenda. How often have you heard, "Studying mathematics will make you think better?" This is an argument which has been used for many years on the beginning geometry student to bribe him into completing the course successfully. Many students rightly suspect that there is more to thinking than proving theorems of somewhat doubtful interest and value to them. The more astute will realize that there is more to geometry. Certainly, by the time one has arrived at college the argument can be advanced and vociferously defended that a student can think well enough without enduring the trauma of mathematics. The notion that mathematics helps you think goes back to Plato (despite the fact that his teacher, Socrates, said that mathematicians couldn't think).

Since the ancient Greeks have been mentioned, we might go on to say that the first liberal arts college was founded by Pythagoras and we all know that Pythagoras was a mathematician. We do not need to have much creative thinking ability to arrive at the role that mathematics played in his school. That was 2600 years ago. What about today?

There cannot be a single answer to that question for each of you either individually or collectively. It appears to be one which you must answer for yourself. It may be that you will answer that mathematics and its study hold no value for you. If so, we are sorry, because we do not think that this is the case. On the other hand, it may be that you will decide that there is something in mathematics for everyone. Now we rejoice with you.

However, before you can decide, you must have some knowledge upon which to base your choice.

Genesis

There comes a time in the intellectual life of every human being when he must ask certain questions about the world in which he lives and his faculty for understanding it. Nearly every philosophical and religious system is dedicated to this problem. Moreover, you need to be aware of the fact that this is one of the ends of any true educational experience.

Answers to these questions come in various forms. However, the content suggests that man coexists in two essentially different worlds. Man has both an inner world and an outer world. Some grant certain correspondences between them, and many are willing to conclude that the way one understands the outer world is, to a large extent, governed by the way we think about things and our understanding of past experiences. Plato, of course, makes this perfectly clear in the beginning of Book VII of *The Republic.* It should also be obvious that many times the way we think and understand is influenced by our experiences. Hence our inner world is, to a large extent, an image of the outer world.

Useful as these answers are, they fail to account for many observable situations. B. I. Whorf* has suggested an entirely different set of relationships as answers to the perennial problem of dealing with reality. The next few pages are a modification of the so-called Whorfian hypothesis. In very simple terms, the Whorfian hypothesis

* B. I. Whorf, *Collected Papers on Metalinguistics.* Washington: Foreign Service Institute, Dept. of State, 1952.

states that the commonly held notion that the mental activities of all human beings are based on a common logic system independent of and prior to language is false. Whorf says that linguistic patterns govern not only how a person perceives an event but also how he organizes this event into concept patterns. Hence, because of the variety of linguistic expression, there must be quite a variety of world views.

As an exercise, develop the idea that linguistic forms found in the ghetto (that is, lack of middle-class forms) provide serious detriment to learning of traditional public-school subject matter.

Exodus

The notion that man resides in an inner and outer world seems obvious. However, we assume that we know which world we are dealing with in a certain instance. But, as we demonstrated above, this is not always the case. In fact, we are not even sure if we might not live in a third or a fourth world, or, perhaps, in many more. Therefore, let us set some verbal forms and see what kinds of concrete ideas result. First, we must free ourselves of the common interpretation of the words to be used. (See I. Kant and H. Dumpty on the nature of definition in the selected readings book which is meant to accompany this text, pp. 78–98.)

Let us suppose that man lives in at least three worlds. We shall call these three worlds *Thought*, *Experience*, and *Language*. Our purpose in the next few pages will be to explore these worlds. However, please do not consider these descriptions complete or completely original.

It is sometimes hard for a man to distinguish when he is in one world and not another. As in the above case of thought and experience, difficulty exists in the case of language. Sometimes language seems to be part of thought. Sometimes it appears as a part of experience. Sometimes it appears as a part of both, but sometimes as part of neither. However, for the sake of clarity, we shall assume three separate worlds and label them *manifolds of existence*. In spite of the fact that we are considering the manifolds as separate, it is nevertheless clear that there are correspondences that can be established among the manifolds. These correspondences are suggested by our previous remarks.

Now let us examine these manifolds together so that their internal structures and the correspondences can be more clearly seen.

Thought

The manifold of thought consists of mental activities. Some of these appear to be quite random in nature while others appear to have taken on special significance for the individual. Those of the latter type are called *thought systems.* Thought systems can range all the way from loosely organized and barely recognizable patterns to very systematized, easily recognized systems.

However, all thought systems, no matter how poorly or how well organized, consist of a more basic unit called a *concept.* Since concepts themselves can be in various stages of development, we single out some of those as being of particular importance. These basic concepts are called *propositions.* Thus, for our beginning purposes the basic unit of a thought system is a particular type of concept called a proposition.

Experience

Let us now turn our attention briefly to the manifold of *experience.* The situation is similar to that of thought. Here the submanifolds are the data of perception. These are called *realms of sensation.* Of course, a realm of sensation is composed of *sensa-*

tions. Finally, the basic type of sensation will be called a *state* of a sensation. We shall assume tacitly for the moment that a state may either be or not be the case. In either case we can have knowledge of the occurrence or non-occurrence of the sensation in question. (Actually, there are a number of difficulties in maintaining this position.)

This hypothesis to some extent underlies much empirical science. Hence, those states which are the case are called *facts*. Those states which are *not* the case are called *counterfacts*.*

Language

Let us remember B. Whorf who maintains that our view of two manifolds must be enlarged to at least three. Whorf makes it quite clear that not only is the way we think structured by what happens perceptually and vice versa, but that what we think is structured by the nature of the language we speak. Moreover, he believes that the way our language is structured is in part responsible for the way we experience things. Finally, a corollary to this is that perception must, therefore, influence language structure.

Since the inner structure of each of the previous manifolds has been presented, let us do the same for the manifold of language. One of the interesting features of this manifold is the observation that an individual has at his disposal more than just one language. Choosing any of the languages at random, one can see that languages are composed of *expressions*. Of all the expressions available some of these are distinguished by being labeled declarative sentences, or simply *sentences*.

---

* For relevant outside material see E. W. Beth, *Mathematical Thought*. Dondrecht, Holland: D. Reidel Publishing Co., 1965, particularly Chapters 6 and 7. See also A. Tarski, *Logic, Semantics, Mathematics*. Oxford University Press, 1956.

Exercises

1. Make a conjecture based only on the material presented up until now as to how mathematics fits into this material. Discuss this with some other individual who has read the material.
2. Explain how language could form a bridge between the manifolds of thought and experience.
3. Make a list of technical terms used so far which are (a) defined; (b) not defined but whose usage is specified by context.
4. What are some of the underlying assumptions of the material discussed so far? Do you consider them to be "self-evident"? If not, what if anything justifies them?
5. What conclusions have been drawn?
6. Explain or cite instances of (a) language independent of thought and experience; (b) thought independent of language and experience; (c) experience independent of language and thought.

Morphisms

In previous sections we postulated that man abides in at least three manifolds—Thought, Experience, and Language. These manifolds are interrelated by certain correspondences. Because of the particular nature of these correspondences, they shall be called *Morphisms*.

Digression

In order to determine how morphisms are established, let us digress briefly. How do men communicate? Each of you can think of dozens of ways. Did you think of a nod, a shake of the head, a wink, a drumbeat in the jungle? On the other hand, did you consider the changing of a traffic light or some music that reminded you of an event in the past? Do not forget the postures and movements of the ballet.

Thus we can see that communication includes all the procedures by which one mind can affect another. However, there seem to be at least three levels of problems associated with communication. These are (1) technical, (2) semantical, and (3) influential. Technical problems have to do generally with the hardware, or the devices, used in communication. They are concerned with accuracy of symbol transmission and reception.

Semantical problems are concerned with the interpretation of meaning by the receiver compared to the intended meaning of the transmitter, or sender. Whereas influence is concerned with the degree of success,the meaning accepted by the receiver leads to desired conduct. The word conduct is to be interpreted rather broadly.

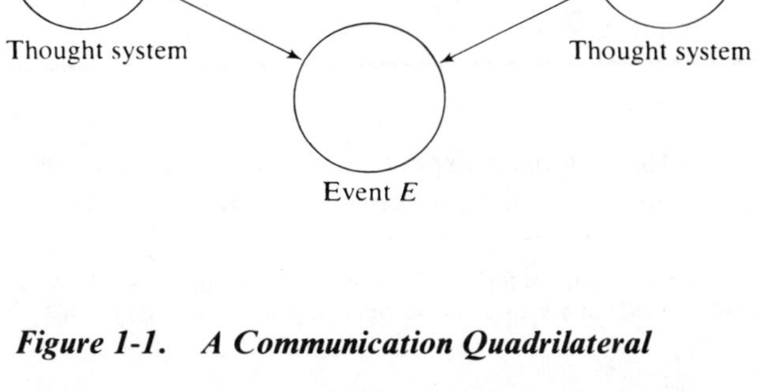

Figure 1-1. A Communication Quadrilateral

To be more specific, suppose Mr. *A* is an observer of some event *E*. (See Fig. 1-1.) Moreover, suppose he wishes to communicate with Miss *B* concerning this event. In other words he has some information to share with, or impart to, her. He obviously expects both understanding and reaction from her. To achieve this, they need "common language." Even if communication is ostensibly possible via a "common language," there may be prohibitions to reception or transmission. Generally, these prohibitions are classed as noise. (See Fig. 1-2.)

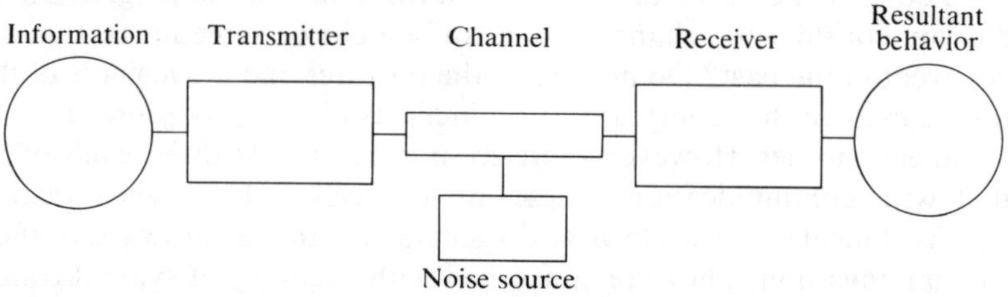

Figure 1-2

It is not our purpose to discuss this particular item in detail. Broadly speaking this subject can be classified as information, or communication, theory. It is related to computers and the way they handle information. However, this is not entirely the case.*

Return

Focusing on the linguistic aspect of our discussion, let us explain why we have placed quotes around the phrase "common language." Earlier, we said that each

* For detailed nontechnical material on information theory, see D. A. Bell, *Intelligent Machines*. New York: Blaisdell, 1962; and G. Raisbeck, *Information Theory*. MIT, 1963.

man has, at his disposal, not a single language, but a number of languages whereby he can communicate certain thought systems and certain experiences. Specifically, a given language is predicated upon a given thought system. However, the mind does not think, and languages do not spring forth of themselves. The mind must think about something, and language must have referents. Thus, out of the totality of all our experience, each language deals with a particular subject matter, or realm of sensation, about which we have formed a thought system.

Thus, there are morphisms established as follows. For every thought system there corresponds a language. Moreover, for every realm of sensation there corresponds a thought system. Clearly, this says that for each realm of sensation a language corresponds. (Why?)

Please recall that we have assumed that thought systems are composed of concepts. Moreover, we have assumed that each language consists of expressions. Clearly, concepts correspond to expressions and vice versa. (See Fig. 1-3.) It is also clear that one can correspond concepts and sensations.

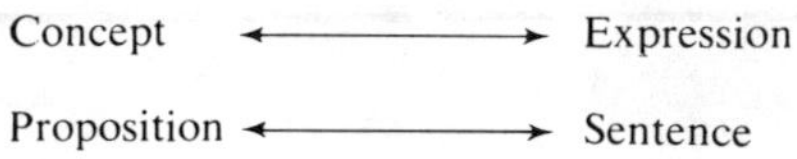

Figure 1-3. Correspondence between a Thought System and a Language

Those particular concepts called propositions correspond to sentences. Hence the smallest unit of language is an expression of a basic unit of thought. This observation recalls the familiar definition of a simple declarative sentence as being the smallest unit of language that expresses a complete thought. Hence, a sentence can be a linguistic expression of a proposition. However, not all sentences or even all declarative sentences are used to express propositions. The only sentences in a language we shall be interested in will be just those that do express propositions.

Turning our attention now to the third manifold, we observed that experience consists of several submanifolds. Since the basic unit of experience is given as a state, one would, in light of the above discussion, suspect that a state corresponds to a sentence which in turn corresponds to a proposition.

I shall leave it to the reader to establish other possible morphisms. These can be assembled in a chart. (See Fig. 1-4.)

Thought		Language		Experience
A thought system	⟷	A language	⟷	A realm of sensations
Concepts	⟷	Expressions	⟷	Sensations
Propositions	⟷	Sentences	⟷	State of sensations

Figure 1-4

Exercises

1. We have said that for every thought system there corresponds a language and for every realm of sensations there corresponds a thought system. Defend or oppose the position taken above that for every subject matter there corresponds a language.
2. What, if anything can be said about possible inverse correspondences? Carefully define these correspondences. Then see if you can find a realization for the definitions; that is, find something to fit the definitions.
3. Comment on the idea that for every concept there corresponds an expression. How about the inverse correspondence?
4. Form an opinion about the possible way in which concepts correspond with sensations and the way sensations might correspond with expressions.
5. Let us agree that a morphism between two manifolds is injective if any part of the first manifold corresponds to at most one part of the second. Are there any injections between manifolds that you can exhibit? If so, what are they? If you can find none, speculate as to why you think there might be none.

Mathematics and the Representation of Truth

A prevailing idea held by many nonmathematicians is the notion that if things are expressed mathematically, then they represent absolute truth. This seems very much a part of the aura of mystery that surrounds the use of mathematics.

Let us explore the question: "What is truth?" Think about your answer for a minute, and then let us consider some of the possibilities. If you have thought about the question at all, you have probably concluded that the question requires qualification. We agree. Further, you have probably concluded that whatever definition at which one might arrive, it must be one of function rather than absolute invariable content. The reason, of course, is because of problems of judgment. (One philosophy teacher used to say that it is far easier to define truth than to judge what is true. We shall find later that a more general statement can be made about most definitions.) Truth, minimally, can be that which works (this is for the pragmatist). On the other hand, truth can be that which is of highest probability. Fans of Sherlock Holmes will remember that he says, "How many times have I told you that when you have eliminated the impossible that what remains, no matter how improbable, is the truth." Thus, according to this view truth is that which is not impossible.

Rather than defining truth, let us provide a base framework that will permit a discussion of truth and what is true. First assume that we have a realm of sensations and a certain state in which a sensation may be. In a given instance the state may or may not be the case. Concretely suppose we consider the situation of whether or not one has a flat tire on one's car. Here we might identify as a realm of sensations all possible sensations concerning a car. The particular sensation concerns the tire and its possible flatness. The state of the sensation is either flat or not flat. Try to exhibit the morphism now indicated and describe the corresponding elements in the other three manifolds. Observe also that normally a state will be presented in the affirmative mood. That is, we will say, "The car has a flat tire" in preference to asserting, "The car does *not* have a flat tire." Of course, either expression represents a state of the tire that can either be or not be.

However, it goes without saying that one must already have a system of thought in mind which sets criteria for determining whether or not a certain sensation should be classified as a flat tire. In other words, there is a thought system with an appropriate proposition which asserts a state. Actually, it seems reasonable to define a proposition to *represent a truth* or to be *true* if and only if it corresponds to a fact and to be *false* if and only if it corresponds to a counterfact. This has led many to define a proposition as the smallest unit of a thought system that can be true or false.

Personally, one may determine truth at the primary level in a relatively easy manner, namely, one can look at the tire, and decide on the basis of physical evidence. A difficulty arises, however, when one tries to communicate the matter to a second party. We can assume that they have a common language. When a person tells you, "I have a flat tire," he knows exactly what he means. But will you know what a flat tire is? More particularly, will you know what he means by a "flat tire"? The best he can do is assume that the term "flat tire" means the same to you as it does to him. (Actually, he may have trouble himself deciding when he has a flat tire.)

Logic

The above type of truth forms a basis for what can be called objective, or experimental, truth—fact and counterfact. However, *our* supposition is that there is another something that can be called truth and there is a way of arriving at it. A product of the cerebral functions and of the individual's patterns of thought, this may be loosely termed *logical truth.*

But what is logic? In the first place we cannot speak of only one logic. To be really correct we need to speak of *a logic* or *logics.* However, each of these things that we term logic does share some common properties. Generally speaking, then, *a logic is a set of rules superimposed on the totality of thoughts in a given thought system that permits one to conclude that certain propositions in a thought system are consequences of certain others.* Before you despair, the following chart will clarify the point:

Thought	Language
Logic	Grammar
Proposition	Sentence
Concept	Expression

A logic provides, therefore, certain rules whereby we can put certain propositions together to form new ones. Logic, being similar to grammar, can be quite intuitive. It can be completely subconscious. It can be so idiosyncratic that the conclusions to which it leads may cause others to call it illogical. On the other hand, a logic can be rigidly defined and agreed upon and used by large numbers of people.

Whether conscious or subconscious, strict or lax, logic provides the individual the rules whereby he can reason, however well.

Theory

Among the many troublesome words in the English language is the word *theory*. The trouble does not stem from the meanings specified by the scientist or the mathematician because these meanings are technical. There are others who, misunderstanding the usage specified by the above group, have allowed certain misconceptions to arise.

In the form in which we wish to view them, theories are thought systems. To some extent all thought systems are theories. Hence, those thought systems we wish to classify as theories must have additional structure. Let us agree minimally that a theory is a thought system with some distinguishing type of structure.

It is certain that many of you are asking why we need theories and why we should bother with learning how to form them and work with them. As you progress through this material we hope that the answer to this will become clear to you and the relevance of theorizing will become obvious. However, as an aid to your understanding we offer the following. Theories are labor-saving devices and in some instances they are life saving. Theories can themselves be an aid to understanding and explaining strings of phenomena, and can be useful in predicting other phenomena. Theories can organize phenomena. As an exercise think of examples of each of these.

If theories are to perform these functions, what properties should they have? Recall that thought systems are composed of propositions. Propositions are those units of thought which can correspond to facts and counterfacts. Thus, for a thought system to be a theory it must provide some way of deciding which propositions correspond to facts *without* the need of verification. In other words a theory might be defined as a thought system equipped with a device to separate the set of all propositions of that thought system into two disjoint subsets, one subset corresponding to fact and the other to counterfact.

However, for reasons that shall be presented later, this turns out to be too much to ask of a theory. We are reminded of what the philosopher Hegel is reported to have said when someone pointed out that some facts did not fit one of his theories: "So much the worse for the facts!"

Let us relax the above requirement. Let S be a thought system; then S is a theory if and only if:

1. S can be divided into two subsets of propositions T and F;
2. T and F have no propositions in common;
3. S has a set of rules governing the obtaining of new propositions from given propositions and their placement in either T or F.

The set T will be called the truths of the theory and the set F the falsehoods. The rules specified by (3) are called the logic of theory. Part (2) says in essence that theories should be consistent. One might hope that a theory would answer all questions asked within it. However, it can be proved that in some cases if we guarantee the answering of all questions, we cannot guarantee consistency.

We cannot leave this discussion without a word about the verification and proof of a theory. Normally, in the sciences and mathematics one does not speak of the proof of a theory, since theories qua theories have no proof. Rather, theories as theories are verified. This means that the propositions which are said to be truths of a theory are tested to determine if they correspond to fact and those taken to be the falsehoods of the theory are tested to determine whether or not they are counterfacts. This testing process is called verification. Thus one test of the goodness, or strength, of a theory is the number of truths which correspond to facts and the number of falsehoods which correspond to counterfacts.

Scientists generally are interested in both the building and verification of theories. Although mathematicians qua mathematicians are generally not interested in the verification of theories, they are interested in the building, developing, and utilization of theories.

Exercises

1. Prepare a short report on the nature of a scientific theory. Relate this idea to the above notion.
2. Ask a number of your friends what they think a theory is. Prepare a composite representing these views. In what way does this represent a theory of theories?
3. Repeat the above exercises replacing "theory" by "truth."

Chapter Two

The Fabric of Historic Mathematics, Part 1

So they went up to the Mock Turtle, who looked at them with large eyes full of tears, but said nothing.

"This here young lady," said the Gryphon, "she wants for to know your history, she do."

"I'll tell it her," said the Mock Turtle in a deep hollow tone, "Sit down both of you, and don't speak a word till I've finished."

Alice in Wonderland

Lewis Carroll

The history of the development of mathematics is an extremely complex subject matter. Because of its nature mathematics is either built upon or is a modification of "older" mathematics. Thus, past mathematical ideas and methods are the experiences that modern mathematicians use in creating new mathematics.

The following materials should not be considered a historical survey for they should be viewed with a good deal of suspicion. They consist mainly of collections of tidbits which are completely unorganized. One cannot fail to note the myths that sustain these surveys.

The linearization of history in these accounts is that which is the most distressing. The past is no more linear than the present. It is a fabric of events of varying complexity and importance woven together and interrelated though, from our perspective, not always rationally.

A complete historical treatment of mathematics would take us too far afield; thus our major aim will be to supply a *framework* for you to view the important events in mathematics. The history of mathematics as we shall see it presented here might be described as two ropes entwined as one. Each rope is itself composed of many strands. One rope is the content of mathematics and the other is the nature of mathematics. Most of you can see that mathematics, like any other branch of knowledge, has increased in content through the years. However, what is probably not so evident is the fact that the very nature of mathematics has undergone a series of evolutionary stages. The character of mathematics has changed since man began to think about and actually do mathematics.

Mathematical Sense

We begin our investigation into the historical development of mathematics by introducing a term—*mathematical sense.* Mathematics undoubtedly arose in the first place because of man's ability to recognize sizes and shapes. For example, primitive man viewing the trees of a forest would have become aware of the concept of vertical. As he began to build houses of some kinds, the ideas of triangle, quadrilateral, and circle would occur to him as idea patterns. Of course, the sun and the moon would have already given him the concept of roundness. On the other hand, as man began to domesticate animals and keep herds, he probably began to actively recognize quantities. Thus, in the early recognition of relative number sizes and in sizes and shapes of spaces and curves, we find the roots of mathematics. Man possessed, subconsciously, what we might call a *mathematical sense* in the form of *geometrical sense* and *numerical sense.* These two, geometrical sense and numerical sense, are two of the most basic kinds of mathematical sense. There is considerable evidence to the effect that these senses exist in various degrees in many lower life forms. Dantzig in his delightful book *The Nature of Number* relates a story about a crow who has a numeric sense in that it can distinguish individual quantities up to four. He also cites an example of a certain species of wasp who senses when it has supplied 20,000 grubs for its young to eat in a single day. All of us are familiar with the apparent geometric sense of bees; they always build their nests in the shape of hexagons.

In the above paragraphs we were not talking about man's investigation of these shapes, sizes, or quantities, but about his apparent ability to recognize shapes and sizes and to compare quantities. Perhaps among the first of the geometrical notions that occurred to early man was that of distance. Traveling from one point to another by different paths and then comparing the times of progress from one point to the other, he might well have concluded that the shortest distance between two points is a "straight" line. Certainly a mind, and not necessarily a very reflective mind, would observe certain outstanding curves in nature that became prominent because of the haphazardness of the other curves around them. The semicircle of the rainbow is an example, or the set of concentric circles formed when a pebble is thrown into water. These would suggest other families of curves. Spirals could have been suggested to him by the shell of the nautilus or by a pine cone. Ideas such as vertical, parallel, and perpendicular would be suggested by any number of natural forms. Without ever putting a name on anything, we see that man could actually amass a large store of mathematical knowledge.

Study of primitive and extinct languages has strongly suggested that the process of counting, which today we take as being among the simplest of all mathematical concepts, passed through many stages before it reached a level that permitted any kind of systemization. Counting ability and an organized way of representing this ability is a necessary condition for any kind of systematized mathematics. Early writing also suggests that a tortuous mental endeavor was necessary to devise a system of numerals which were freed from the concrete objects they might be used to represent.

Inductive Mathematics

Later, man began to have enough time to reflect. His sole problem was no longer survival and he was able to think consciously about other matters. He began to actively search for relationships, and at this time we find a new type of mathematics emerging. This mathematics, or this phase of mathematics, we shall call *inductive* mathematics. Another term that might be applied here is that of scientific mathematics. (We use the word "scientific" here to mean "based on experiment.") The human intelligence had reached a level where it could examine a set of relationships and extract from them a generalization. We find this mode of thinking within the frame of highly developed urban civilizations and well-formed economic processes.

Let us now consider some examples of experimental mathematics. Suppose a farmer wishes to enclose a rectangular field. In order to supply water, he wishes to have one side of the field bounded by a river. Referring to Figure 2-1, we see how this

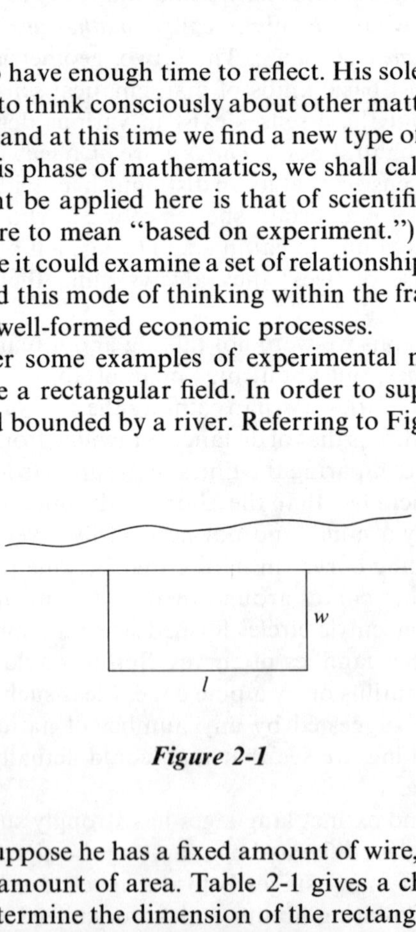

Figure 2-1

must look. Let us also suppose he has a fixed amount of wire, say 200 feet, with which to enclose a maximum amount of area. Table 2-1 gives a chart of experiments that may be performed to determine the dimension of the rectangle of maximum area and fixed parameter. The table now suggests that the dimension of the field should be 50 feet by 100 feet. Now an observant individual might wonder if there is any significance in the fact that 50 is one-half of 100. As an exercise, try a different perimeter and see if there is any significance to the fact that the width is one-half the length.

Table 2-1

Width	Length	Area
10	180	1800
20	160	3200
30	140	4200
40	120	4800
50	100	5000
60	80	4800
70	60	4200
80	40	3200
90	20	1800

In this example you can see certain processes in action. First there is the process of experimentation. Once a series of experiments has been made there is the attempt to draw some sort of general conclusion. The latter process—the deriving of a general conclusion from a number of specific instances—is known as *inductive* reasoning. Today we would not let mathematics be based solely on this type of reasoning, for it does not constitute "proof." However, we still use inductive reasoning when we verify theories or suggest new ones. Indeed, in the above example we have formulated a theory or at least laid the foundation for formulating one. (What is it?)

Babylon et al

No historical account can be complete without mentioning some dates. When was it that mathematical sense ceased to dominate and was replaced by experimental mathematics? Of course, mathematical sense is still with us; without it we could do nothing mathematical. So the proper question should be when did man begin to be experimental in his mathematics? It seems that no matter how far back in recorded history we go, we find evidence of some type of experimental mathematics. Hence, it appears that this mathematics predates recorded history and goes back farther than 5000 B.C.

Among the many great civilizations of antiquity, four are important to our discussion: Egypt, Babylon, Greece, and Rome. This is not to say that other civilizations did not contribute to mathematics; however, numerous events make it difficult to determine their actual influence. Of the four civilizations mentioned above, two—Greece and Babylon—seemed to have contributed a majority of mathematics. The other two—Egypt and Rome—never progressed beyond elementary practical arithmetic and mensuration. This is true in spite of the large amount of publicity Egypt and Rome have received to the contrary.

Priority in significant mathematics of this period seems to belong to Babylonia, where numerical, algebraic, and geometric methods existed at least as early as 1700 B.C. Whereas most of our information about Egypt comes from two papyri, one the so-called Rhind papyrus (written about 1650 B.C.) and the Moscow papyrus (written about 1900 B.C.), neither of the two Egyptian sources is of the excellent quality of mathematics found in the Babylonian sources.*

The Birth of Axiomatics

For many hundreds of years myths and legends about Thales of Ionia and Pythagoras of southern Italy have dominated the minds of mathematics students. In actual fact, we have little evidence in the form of primary sources to substantiate the marvelous claims made concerning these individuals. Most of our information comes from secondary sources such as reports and suggestions of thoughts written by philosophers and historians of a later period. Thus the first three- to four-hundred years of Greek mathematics we know only by inference, speculation, and hearsay.

However, we do have some fairly good secondary sources. Among the most important of these is a work by Proclus (A.D. 500; note that this is 1000 years after the

---

* For a complete discussion of the content of these sources, see O. Neugebauer, *The Exact Sciences in Antiquity*. Providence, R.I.: Brown University Press, 1957.

period discussed) entitled *Commentary on Euclid, Book 1*. Proclus apparently had at his disposal documents no longer available to us. Among these documents was one written by Eudemus, a student of Aristotle. Eudemus had written a comprehensive history of Greek mathematics encompassing the span 600 B.C. to 335 B.C. The first few pages of Proclus are devoted to a summary of Eudemus' work and have consequently been termed as the "Eudemean Summary."

According to the Eudemean Summary, Greek mathematics began with Thales who lived around 600 B.C. Thales is credited with being one of the seven wise men of antiquity, and so he gave an air of respectability to mathematics. It is reported that Thales had the idea that mathematical results should be based not on the inductive process (as outlined in a previous section of this work), but upon the process of *deduction*. The deductive process depends not upon specific cases suggesting or leading to generalities, but upon beginning with general statements and producing specifics by small chains of reasoning. This gives one a type of certainty not to be found in depending solely upon experimental verification.

Let us consider an example of the sort of thing with which Thales is credited:

> ***Theorem:*** If two straight lines intersect, the pair of vertical angles A and C are equal. (See Figure 2-2.)

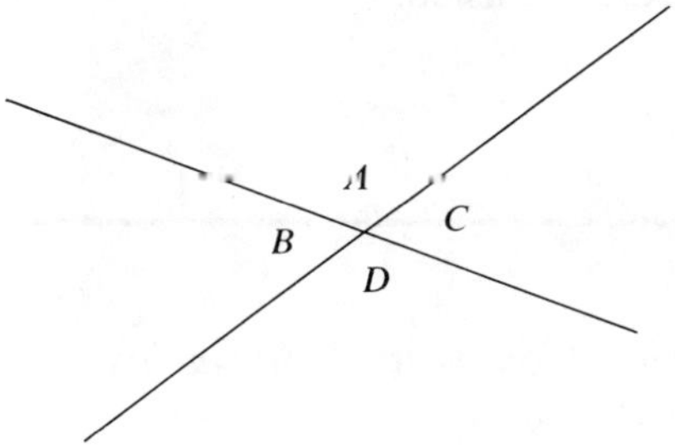

Figure 2-2

> ***Proof:*** Let us prove angle A equal to angle C. We note that angle B and angle C together form a straight angle. We also note that angle A with angle B and angle C with angle B form equal quantities. Thus taking away angle B results in A and C being equal.

Admittedly, this is a simple result proved in a simple way; moreover, this result was known to the Babylonians long before Thales. There is none of the sophistication one would expect of a learned mathematician. However, we must judge Thales or whoever made this move not by the result or by the "proof" but by the way the problem was approached and ultimately solved. (The importance of the approach to a problem and how it is solved lies at the core of the "New Math.")

The Eudemean Summary credits Pythagoras as being the next man to continue the purification of mathematics. Pythagoras and his mystical Pythagorean Brotherhood are so surrounded with lore that we can be sure of little that is associated with his name. Pythagoras was some fifty years younger than Thales and may have studied with him. The Brotherhood, which reflected many of the mystical teachings that Pythagoras had somehow acquired, took the study of mathematics as one of its major tasks. This group lasted for over two hundred years, and in this time produced much mathematics by deduction. More than likely they took most of the known experimental mathematics and attempted proofs of it.

In this way the deductive chains became longer and longer until someone decided that it might not be a bad idea to assemble all of mathematics into one long chain. This was first done presumably by Hippocrates of Chios (440 B.C.). Members of Plato's Academy, Leon and Theudias, made a similar attempt in the first half of the fourth century B.C. All of these led up to the best-known effort by Euclid in 300 B.C. In the *Elements* of Euclid there is one long chain of 465 links, wherein all of known mathematics was supposed to be listed.

Concrete Axiomatics

Let us now examine some of the features of Greek mathematics as it was exhibited by Euclid. As we mentioned above, his major work was entitled *Elements*. It was a treatise on the foundations of mathematics as well as on geometry. From it we may infer some of the notions the Greeks held concerning mathematics.

First, Euclid introduces some fundamental ideas. These ideas he attempts to define in terms of physical realities by means of explanatory definitions. These "definitions" play no essential role in the results he later obtains. Thus the primitive terms such as point, line, and plane are in effect "undefined."

Second, Euclid states certain propositions which he does not prove concerning the primitive terms. These the reader is asked to accept as being self-evident relative to the primitive terms and their physical-world meaning.

The main objective in this preparation is to deduce theorems. This is the third item on Euclid's agenda. At the same time he provides the reader with definitions of such terms as triangle, right triangle, and hypotenuse.

The entire process is designed in such a way to convince the reader, by starting with readily granted postulates, of the truth of propositions of which he would be far from convinced had they been asserted at the outset. This is done by a series of simple steps such as outlined above. Thus the Euclidean program of dealing with real-life geometrical material is to give a brief list of "readily observed" phenomena and use them as given items. They are stated using words that are technically undefined. From these items are deduced secondary propositions by a series of simple steps. The truth of these items then is asserted on the basis of the truth of the given items and the correctness of the simple intermediary steps.

This approach can be called *concrete axiomatics.* The term *concrete* is used because the method is applied supposedly to things in concrete reality. The usage of *axiomatics* stems from starting with given items called axioms or postulates. A further discussion of this method will be presented in a later chapter.

A Crisis

The Eudemean Summary suggests that axiomatics arrived on the scene by a slow process of evolution. However, there is considerable evidence to support the notion of revolution. Usually abrupt changes take place because of some crisis. The introduction of axiomatics seems to be a rather abrupt change. Hence, there must have been some early crisis in mathematics, so severe that it brought about a complete change in thinking.

It is known that around 400 B.C. a devastating challenge to the foundation of mathematics did occur. It was the discovery by some Pythagoreans that there existed certain numbers that did not fit the picture of reality that the Pythagoreans had established. The entire Pythagorean philosophy and religion was predicated on a concept of reality based entirely on number. Number for them meant either a whole number or the ratio of two whole numbers which they called rational numbers. Their entire mystical philosophy was tied to the idea that even such a thing as honesty had its basis in whole numbers. Try to imagine the chaos that resulted in the discovery, through a deductive process, of the existence of numbers that were neither whole numbers nor a ratio of whole numbers. These new numbers are called irrational, meaning not rational in the sense given above.

Let us see how this discovery may have come about. Suppose we are given a right triangle ABC as shown in Figure 2-3 with sides one unit each. Then because the triangle

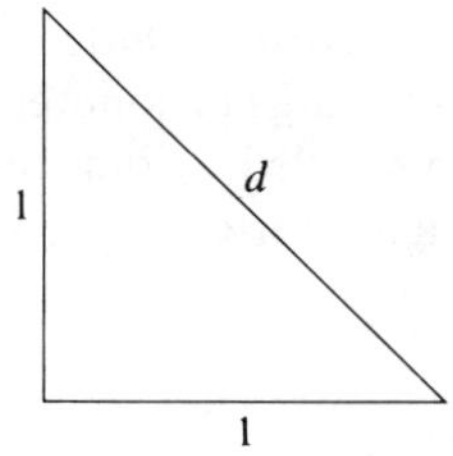

Figure 2-3

is a right triangle, $a^2 = 1^2 + 1^2 = 2$. We shall now show that a is not rational. Obviously a is not a whole number, thus it must be the ratio of two whole numbers. In other words $a = p/q$ where p and q are whole numbers and p/q is reduced to lowest terms. This says $a^2 = p^2/q^2$ or $2 = p^2/q^2$ and $2q^2 = p^2$. Thus p^2 must be an even number, from which it follows that p is an even number. (Take time to convince yourself of this before going on.)

We may therefore write

$$p = 2k$$

for some whole number k. Hence

$$p^2 = 4k^2 = 2q^2,$$

or

$$q^2 = 2k^2.$$

From this it follows that q^2 and hence q must be even. Note what we have now established. We assumed that both p and q had no common factors. Then, we were able to show that they did have a common factor. (What is it?) Logic now dictates that there cannot be a rational number with these properties. Thus $a = \sqrt{2}$ is not a rational number.

What are the consequences of this result? It had always *seemed* reasonable that if two line segments were drawn, then there must be some common unit of measurement that will divide each line segment equally. Granted the unit might have to be very small, but surely there must be one.

We shall now show that there exist two line segments for which it is *impossible* to find such a unit. (Please note the use of the word "impossible." Refer to the Halmos article.) Consider the square $ABCD$ with diagonal AC. Suppose further that the length of each side is a. Then $\overline{AC}^2 = a^2 + a^2$ or $\overline{AC}^2 = 2a^2$. Thus $\overline{AC} = a\sqrt{2}$. Suppose now that AC and a have a common unit called c. Then c divides $a\sqrt{2}$ a whole number of times and c divides a a whole number of times. Thus $a\sqrt{2} = cm$ for some whole number m and $a = cn$ for some whole number n. Thus $cn\sqrt{2} = cm$ or $n\sqrt{2} = m$ for some whole number n and m. This says $\sqrt{2}$ is rational which is a contradiction to what we have previously shown.

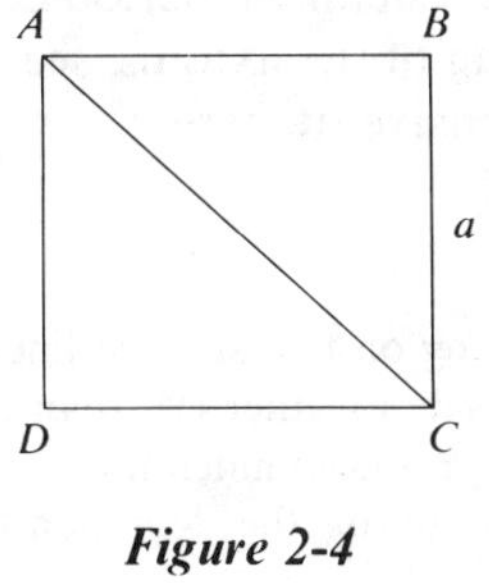

Figure 2-4

Thus, there exist pairs of line segments with no common unit. These we call *incommensurable*. Those with common units are called *commensurable*.

If you have followed the tedium of the preceding arguments, many of you are probably asking the same question you may have asked in high school: "So what?" The answer is simply that the whole Pythagorean treatment of similar figures and proportionality was based on every pair of lines being commensurable. In one simple theorem a lifetime of work was demolished, all because there was no careful listing of basic assumptions at the start. The solution to the above difficulty was experimental axiomatics.

Dissatisfaction with Euclid

While many problems were solved with the advent of experimental axiomatics, others still remained. Even though some excellent mathematics was done after the time of Euclid, there was no further development in axiomatics until the late nineteenth and early twentieth century. At this time two events conspired to bring about another revolution in the nature of mathematics. One of these was geometric and the other was algebraic. We shall concern ourselves first with the geometric.

Even though nothing substantial was settled before 1800, some dissatisfaction existed among the Greeks of the Euclidean period concerning some of the axioms of geometry. (For a complete listing of the axioms, see Chapter 3.) It seems that one of the axioms was particularly offensive. It states:

> If a straight line falling on two straight lines makes the interior angles on the same side together less than two right angles, the two straight lines, if produced indefinitely, will meet on that side on which the angles are together less than two right angles.

If you examine the rest of the postulates, you will find that they are readily understood and acceptable. However, we now need to examine Figure 2-5. Line k is the straight line falling on the two straight lines *m* and *n*, making interior angles 1 and 2 whose sum is less than two right angles. The axiom thus says that *m* and *n* when extended indefinitely will meet on the side where 1 and 2 are.

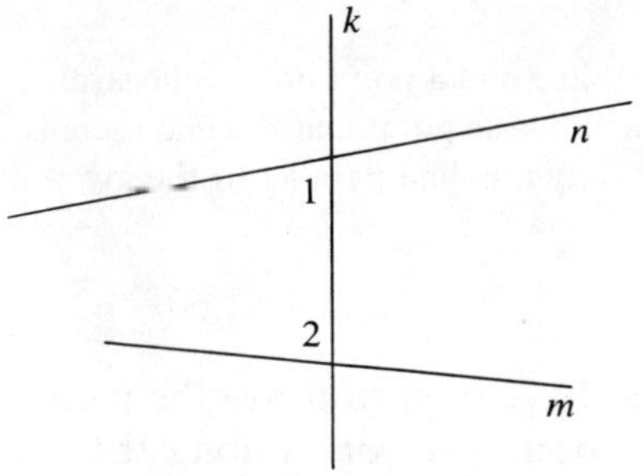

Figure 2-5

In passing, we note that prior to Euclid's attempt, the theory of parallels had always provoked controversy. (Aristotle discusses it in his writings.) It seems the problem was that of circularity: assuming what one wants to prove. Euclid at least avoids this.

One might infer that Euclid himself was unhappy with his troublesome axiom for it is not until proposition 29 of Book 1 that he uses it. It is interesting to note that the converse of this axiom, formed by interchanging hypotheses with conclusion in the axiom, is Proposition 17. These are some of the things that astounded mathematicians:

1. The axiom was believable as a theorem;
2. The converse of the axiom was provable as a theorem;
3. Euclid included the proposition as an axiom with the tacit assumption that it could not be proved.

What can one do if one is unhappy about a postulate? Basically, aside from ignoring it, which is difficult, one can do one of two things. One can try to replace it with something that is acceptable or try to deduce it as a theorem from the remaining postulates. Both were tried in the present case. Among the most popular of the replacement postulates is one by John Playfair (1748–1819). It is one with which most high school students are familiar instead of Euclid's original. It reads:

> Given a straight line and a point not on the straight line, in the plane determined by the point and the line there is through that point one and only one line parallel to the given line.

Many are the ones who have tried to prove the parallel postulate as a theorem. All of these attempts have been in error. Among the most common errors is the *implicit* assumption that straight lines can be continued in both directions indefinitely, but so can an arc of a great circle on a sphere.

Finally, mathematicians quit trying so hard and the idea was conceived that perhaps there was a simpler explanation of the whole affair. In this way, we have the advent of non-Euclidean geometries. Many men contributed to this occasion. We shall just mention some of them: Girolamo Saccheri (1667–1733), Johann Lambert (1728–1777), Adrean-Marie Legendre (1752–1833), Johann Bolyai (1802–1860), Nicolai Ivanovitch Lobachevsky (1793–1856), and Bernard Riemann (1826–1866).

Even though non-Euclidean geometry is important in its own right, its implications are of far more interest. First, it was shown by Beltrami and others that the geometry of Lobachevsky was self-consistent. Moreover, it can be shown that the parallel postulate is independent of all the other postulates of Euclid and this with the consistency of the Euclidean postulates shows that we cannot hope to deduce the parallel axiom from the others. More important is the liberation of geometry from the traditional mold. It now becomes apparent that, for the mathematician, the postulates of geometry are assumptions whose physical truth or falsity need not be of concern. Consistency is the guiding light. The mathematician may take any axioms he pleases so long as they are consistent with each other. Thus, *belief on the part of the reader is no longer of immediate concern.*

Mathematics thus emerges from this emancipation as an arbitrary creation of the human mind. The mathematics of a mathematician is now seen to be related to the real world in much the way the character and events of a novel are related to real-life situations. Thus the novelist and mathematician have much in common.

Form Freed of Necessary Content

As mentioned above, mathematicians no longer had to concern themselves with the necessity of conforming to physical reality. Nor did the axioms that they chose need to possess the quality of instant belief. We must not, however, think of the new mathematics as being a useless figment of someone's imagination. There are physical studies where such things as non-Euclidean geometry play an essential role. Einstein, a physicist, found a geometry of Riemann quite useful in discussing general relativity. It is interesting to note that the space we see is probably best described by Lobachevskian geometry.

Most important to mathematicians and to mathematics generally was the new perspective. Form, not content, was the guide. In a sense, this really brings us back to the Greeks and in particular to Plato (428–348 B.C.) who, though probably not a mathematician, contributed extensively to mathematics. (See Plato's articles in Readings.)

As can be inferred from many instances in his dialogues, Plato believed that form was the only reality and that knowledge of mathematics would provide one with knowledge of form and hence reality. Here in the nineteenth century, the mathematicians are saying that mathematics deals exclusively with form.

How then does this affect axiomatics? The answer is quite clear. Undefined terms are chosen for convenience. The axioms are also chosen for convenience. They are chosen to dictate the usage of the undefined terms. Definitions are made so as to call attention to important concepts and to name them. Finally, it is hoped that theorems can be deduced. Their correctness is judged strictly on the basis of formal deduction as dictated by the rules of a logic. Content is of no concern. Thus a new pattern of axiomatics had arrived on the geometric horizon. This pattern we shall call *formal axiomatics.*

The Revolution Comes to Algebra

Up until the beginning of the nineteenth century, mathematicians regarded algebra as just symbolized arithmetic. Indeed in many respects it is still thought to be that by many students and teachers alike. In arithmetic one uses specific numbers and operates on them according to certain rules. In algebra, according to this view, these numbers are replaced by letters representing numbers. One does the same operations with the letters as one does with numbers.

Fortunately, a set of British algebraists began to formulate the idea of an *algebraic structure*. George Peacock (1791–1858) in 1830 seems to have been the first to express this idea. Ten years later Duncan Gregory (1813–1844) spelled out the properties of commutativity and distributivity of operations. Probably the most important event was the discovery of an algebra that did not always behave as numbers did.

Since at this time algebra was considered to be arithmetic generalized, no one had ever thought to ask if it were possible to construct an algebra in which commutativity for "multiplication" does not hold. (For an easy example of an algebra of this type, see Denbow and Goedicke's article "Group Theory" in the Readings.)

Thus algebra experienced the liberation that geometry did. Algebra is no longer tied to numbers. Indeed, it now can be used to investigate problems in geometry and it is subject to formal axiomatics.

Abel and Galois

Even though algebra can in some sense be considered as generalized arithmetic, it is also an algebraic tradition to be concerned with the solution of equations. Among the most important classes of equations that have been studied extensively are those which are called polynomial equations. They take the form

$$a_n x^n + a_{n-1} x^{n-1} + \cdots + a_1 x + a_0 = 0.$$

The a's with their subscripts are usually real or complex numbers and x is usually taken as "unknown." The object of the usual exercise is to find all those values of x which make the expression on the left-hand side of the equal sign zero. If an x can be found which does that, that x is said to be a solution of the equation. No doubt all of you can solve an equation such as

$$2x + 3 = 0$$

and find that if $x = -(3/2)$, then the left side is zero. (Check it.) Many of you can probably solve

$$x^2 + 3x + 2 = 0$$

to discover that $x = -1$ and $x = -2$. Hopefully, many of you will be able to obtain solutions for

$$2x^2 - 4x + 3 = 0.$$

In this case the solution is much more difficult to obtain. However, it can be shown that the following values for x are solutions

$$x = \frac{2 + \sqrt{-2}}{2}$$

and

$$x = \frac{2 - \sqrt{-2}}{2}$$

These solutions were obtained by use of the famous quadratic formula which produces solutions to the general equation

$$ax^2 + bx + c = 0.$$

The quadratic formula states that

$$x = \frac{-b + \sqrt{b^2 - 4ac}}{2a}$$

and

$$x = \frac{-b - \sqrt{b^2 - 4ac}}{2a}.$$

By about 1500, mathematicians knew these algebraic methods for solving first- and second-degree polynomial equations. They knew also algebraic and geometric methods for solving third- and fourth-degree equations; that is

$$ax^3 + bx^2 + cx + d = 0$$

and

$$ax^4 + bx^3 + cx^2 + dx + e = 0.$$

However, an equation such as

$$ax^5 + bx^4 + cx^3 + dx^2 + ex + f = 0$$

absolutely defied them. Until the first of the nineteenth century, mathematicians generally felt that someone would eventually find an algebraic solution. By the way, just because no algebraic method of solution existed at that time does not mean that other methods of solving problems were not available. For example, using the methods of his newly invented calculus, Newton had a nonalgebraic solution.

The hope of a general algebraic solution was completely demolished by two very young mathematicians, who had the misfortune of dying before they were thirty years of age, leaving future mathematicians plenty of work to do based on their ideas. The first of these was Niels Henrik Abel (1802–1829) of Norway. The other was Evariste Galois (1811–1832) of France. Both men tried to solve the fifth-degree polynomial equation; both thought they had solved the problem only to discover an error in their reasoning. Abel, though, did show that the general fifth-degree equation had no algebraic solution. To do so he used some group theory. These ideas were just beginning to make their appearance in geometry as transformation groups. For a very elementary treatment of a transformation or permutation group, see Denbow and Goedicke's article in the readings.

Making use also of groups, Galois was able to show that there was no general algebraic solution to the fifth-degree polynomial equation. He was able to then find necessary and sufficient conditions that any polynomial equation have an algebraic solution. This involves associating a group with each polynomial equation. He could then tell about the solvability of the equation by examining properties of the group. Thus some geometric ideas came to the rescue of algebra.

Mathematics thus entered a second golden age not unlike the one which flourished in early Greece. It was not to last for a new breed of mathematicians was venturing forth. Before we can study the consequences of these events, we must break and study axiom systems in more detail.

THE WHOLE IS THE SUM OF THE PARTS

Chapter Three

Axiomatics

The first person he met was Rabbit.
"Hallo, Rabbit," he said, "is that you?"
"Let's pretend it isn't," said Rabbit,
"and see what happens."

Exploration to North Pole
Winnie-the-Pooh

A. A. Milne

Why Axioms?

At the end of the last chapter we introduced the term "axiomatic method." This term and certain related areas are now our center of interest. At the heart of the axiomatic method is that which we call an axiom. In the beginning the word axiom will be used to mean an underlying assumption, a proposition that is assumed to represent a factual state.

However, this was not always quite the meaning of the term axiom. For the Greeks who first conceived the notion of needing unproved statements, the term had a slightly different meaning. In fact, the Greeks had two words, axiom and postulate, which they used to distinguish two types of assumptions. For Euclid and other Greeks, the notion of axiom seems to indicate a basic assumption that is common to all subjects. These he listed as:

1. Things which are equal to the same thing are equal to each other.
2. If equals be added to equals, the wholes are equal.
3. If equals be subtracted from equals, the remainders are equal.
4. Things which coincide with one another are equal to one another.
5. The whole is greater than the part.

On the other hand, a postulate was for Euclid a basic assumption that is peculiar to the subject at hand. Pertaining to his geometry he lists the following assumptions as postulates:

1. A straight line can be drawn from any point to any other point.
2. A finite straight line can be produced continuously in a straight line.
3. A circle may be described with any center and any distance.

4. All right angles are equal to one another.
5. If a straight line falling on two straight lines makes the interior angles on the same side together less than two right angles, the two straight lines if produced indefinitely will meet on that side on which the angles are together less than two right angles.

The Greeks, recall, decided that to reason correctly one must start with general statements and proceed according to certain rules to derive particular instances of these generalities. But, where do we get these generalities? One answer might be that they can be deduced from other generalities. This answer, unfortunately, is unsatisfactory since one might ask the same question of the general generalities. This question can be repeated as often as one pleases without end.

Hence, we have the search for general self-evident truths. The approach then was to find a suitable set of statements everyone could agree upon as representing self-evident truth. From these a minimal list was chosen and was called a set of axioms and postulates. The Greeks felt that there were such things as general statements, and that these were self-evident truths. One question, of course, that we could ask the Greeks is, "What constitutes a self-evident truth?" Now we know from where we stand that self-evident truths are hard to come by, because one can always ask, "Self-evident to whom?" What we might claim to be self-evident, someone else may not. Who is to judge the correct one? *Naturally, a self-evident truth is one with which everyone agrees!* How many things do you know with which everyone agrees? There are probably not very many.

Undefined

This difficulty of trying to prove everything convinced Euclid and his contemporaries of the impossibility of proving every statement. However, they seem to have overlooked the fact that similar reasoning can be applied to the definition of words.

To review, we observe that in order to prove a statement, certain secondary statements must be used. To prove these secondary statements, certain tertiary statements must be used. Either this process continues indefinitely, which is unlikely though possible, or the reasoning becomes circular. The latter means that we find ourselves proving something in terms of itself. Neither situation is acceptable. The alternative is to state that something cannot be proved. So, with our superb insight (hindsight) we say to the Greeks, "If one wishes to define a word, one must use secondary words, etc." Here the situation is far simpler, since at a given time in a given language we have at most a finite number of words available for use. Hence, at some stage we will have defined the given word in terms of itself.

So these remarks lead us to say that if one attempts to treat any subject matter systematically, one is inevitably led to base the treatment on some unproved statements and some undefined words. One might ask, "How can we reason about undefined terms? Didn't Plato caution us to define all our terms?" The answer is that we know the usage of our undefined terms by what we assume concerning them in our axioms. It therefore follows that definitions made in our system must ultimately rest on undefined terms.

In summary, by an axiomatic method we mean the use of something called an axiomatic system. The latter can be defined as a collection of statements beginning with some of the statements being labeled axioms. The axioms involve the use of both defined and undefined words. The remaining statements are either explicit definitions or provable statements. The latter are generally called theorems. Consequently, the axiomatic method consists of the deduction of theorems from a given set of axioms, undefined terms, defined terms, and previously deduced theorems.

Experimental Axiomatics

There are two types of axiomatic systems as was mentioned in the previous chapter. The first of these will now be called experiential systems. As the name implies, experiential axiomatics is to a large extent concerned with objects of experience. The symbols, or words, used have a supposed real content. Such symbols might be "house," "mother," "pain," or "cabbage." Any symbol which tells the reader something relating to his experience must at first glance be classified as experiential and these symbols when axiomatically treated evolve into experiential axiomatics.

Suppose now that we have some concrete topic which we wish to develop by an axiomatic procedure. As an elementary example let us take an axiomatic approach to carpentry. We realize that we will have to converse about a number of technical terms such as hammer, nail, board, wood, beam, and other things of this nature. However, as we observed above, any technical term must be defined using other technical terms. Of course, we now know the difficulties this procedure produces. Hence we might select a few basic words with which we assume our audience is already familiar. We might, as a gesture of helpfulness, offer some brief explanatory information concerning these words, but no definition. In the example of carpentry we might start with such words as hammer, wood, and nail. Various kinds of hammers could consequently be defined. A similar procedure applies to wood and nail.

It must be mentioned that any treatment such as we are presenting will in all probability involve the use of nontechnical words whose meaning will be assumed. This set of words consists of words such as "be," "have," "is," "such," and "that." By themselves their meaning is not crucial. Besides, they are common everyday words that everyone understands by usage, are they not?

Recall now that the object of any systematic discussion is to deduce new statements from old statements. However, we have seen that not all statements are deducible unless we allow circularity of reasoning which to most people seems unacceptable. Hence in order to avoid the situation of deducing A from B and then later deducing B from A, one should start by assuming certain statements from the beginning. These we have called axioms. Thus we might assume:

Axiom C1: There is at least one hammer, one nail, and one piece of wood.

Axiom C2: For every piece of wood, there exists at least one nail and one hammer such that the hammer puts the nail into the wood.

Axiom C3: Every nail joins exactly two pieces of wood.

At this time we might wish to supply the reader with a definition.

Definition C4: The process of putting a nail into a piece of wood using a hammer is called hammering.

Finally, we state the following axiom.

Axiom C5: For every piece of wood, there exists at least one other piece of wood that can be joined to it by hammering.

With all of these facts available, it would seem as though a deduction or theorem was in order. One such theorem is the following item.

Theorem C1: Given a pair of pieces of wood there exists a nail that joins them.

I hope that the reader will make the attempt to show that this is a valid deduction. The reader should also see if he can find other possible deductions.

Formal Axiomatics

There is a problem in basing a system of axiomatics on the real world. When people deal with objects with which they are familiar, they want the objects and the deductions to agree with their experience. For example, if offered the axiom "Grandmothers are fat" in a problem, it is very possible that you would become concerned with the relationship of *your* grandmother (who might be thin as a rail) to the axiom. The net result is that you are sidetracked from the discussion of the relationship of certain classes of objects, and end up discussing only one class—grandmothers. Problems of this nature led to the discovery that Euclidean geometry was not the only possible geometry, and that concrete mathematics was not the only possible mathematics. A system was evolved that avoided, or rather, did not include, the problem of experiential axiomatics. This system is called *formal axiomatics*.

Again in the name is the key. Formal axiomatics is concerned with form, and not content. The concept of formal axiomatics is of rather recent origin, the later part of the nineteenth century. David Hilbert is generally credited with the first formulation of formal axiomatics, but other men—Pasch and Peano—also worked on the problem at about the same time. Strangely enough, the idea was a product of an attempt to strengthen Euclidean geometry and to help overcome its difficulties.

After the development of formal axiomatics as Hilbert conceived it, Peano and others tried to do much the same thing with algebra. Peano's first attempt was in trying to discover whether or not arithmetic had a formal axiomatic structure. In other words, if geometry is based upon an axiomatic structure, why not arithmetic? Hence, as an undefined term for his axiom system Peano took things he called natural numbers and an undefined operation called the successor operation. He then assumed the following postulates.

N1: 1 is a natural number.

N2: Every natural number has a unique successor.

N3: No natural number has 1 as its successor.

N4: Distinct natural numbers have distinct successors.

N5: If we have a statement concerning natural numbers which represents a fact for the natural number 1 and if, when the statement represents a fact for a natural number n it also represents a fact for n's successor, then it represents a fact for every natural number.

A twentieth-century philosopher, Bertrand Russell, added the idea of what is called the "propositional function." Since we are modifying Russell's idea somewhat, we shall say *propositional form* instead of propositional function.

Propositional Form

If you will recall the grandmother axiom presented earlier, you will agree that in the light of Chapter 1 it is a proposition. It can be said to represent either a fact or a counterfact. It is a simple declarative sentence. Since grandmothers might be controversial, let us move to a different subject—seasons. Let us for the moment grant that "Spring is a season" is a proposition. Now, of what is this proposition composed? First, there is a noun "Spring." Second is the copula "is," and last is a noun phrase, "a season." If we look back to the grandmother proposition, we will find the same structure. Thus, we can generalize that any proposition of the above type can be divided into three sections, that is, noun, copula, and noun phrase. (We might as well drop the word "phrase.") Before we leave this area, however, we might observe that the copula is always a form of the verb "to be."

Russell observed the structure of the proposition, and concluded that, since nouns are meaningless without previous definition, the structure could be expressed as:

"(Some noun) is (some noun)."

All that you have is that "*noun X* is *noun Y*," or "*X* is *Y*." Thus, we have the propositional form, "X is Y." Formally speaking "*X* is *Y*," "*some noun* is *some noun*," "Spring is a season," and "Grandmother is fat" are the same. The difference between them rests in the fact that the propositional form, having no content, has no *a priori* truth value. The propositional form, having no truth value, is not a proposition. It can, however, become a proposition simply by substituting any noun for *X* and *Y*, respectively.

We can see how the propositional form eliminates the problem discussed in connection with experiential axiomatics. The experiential axioms are reduced or replaced by axiom forms, or axiom schemata as they are sometimes called. The relationships are established as existing between a set of unspecified *X*'s and *Y*'s, and content can be reinserted.

An Example

Before we look at the formal system, let us look at another type of experiential axiomatic system, different from the carpentry example. Let us suppose that there is a collection of objects which shall be called *cats*. Also, suppose that there is a relationship among the cats called *scratches*. Further, assume that this collection of cats, and this relationship that exists among cats for some reason or another satisfies the following axioms:

1. There exists at least one cat.
2. There is a cat that is not scratched.
3. If two cats A and B scratch the same cat C, then A and B are the same cat.
4. Each cat scratches at least one other cat.
5. No cat scratches itself.

We have assumed in Axiom 1 that there is one thing to talk about. This axiom can be called an axiom of existence. In regard to Axiom 2, nearly everyone would agree that there is a cat that is not scratched. Axiom 3 seems somewhat artificial, however, we hope not unduly so. Axiom 4 is certainly not unreasonable. Axiom 5 is important because it tells us that the word "scratches" is to be used in a very special way.

As mentioned above, the objective of axiomatics is to deduce theorems. With this in mind, we will prove the following theorem.

Theorem 1: No cat is scratched by more than one other cat.

Proof: Suppose some cat, call him C for reference, is scratched by two cats A and B, then axiom 3 says A and B are the same. This proves the theorem.

Later in our discussions we shall examine exactly why this is a proof.

Two questions now arise. How many cats does a given cat scratch? How many cats are there in the cat population? Let us attempt to answer the last question first.

To begin, let us introduce some notation that will help in the discussion. This is similar to drawing a picture in geometry. If A and B are cats, and if A scratches B, let us represent this situation by simply writing A s B. By A s′ B, we mean that A does *not* scratch B. Now note that Axiom 1 says that there is a cat that is not scratched; we will call him $C1$. Axiom 4 says $C1$ must scratch some cat; we will call this one $C2$. $C2$ must scratch a cat other than himself and $C1$. (Why?) At this stage notice that one and only one of two alternatives can occur: (1) This process terminates after some finite number of cats, or (2) this process continues indefinitely and leads to the conclusion that there are infinitely many cats. Let us examine the first alternative. To this end suppose there is a last cat Cn; by this we mean that there are a finite number of cats. However, Cn must scratch someone. Clearly, it is false that Cn s $C1$. Recall that $C1$ fulfills Axiom 2. If $C1$ s $C2$, then Cn s′ $C2$. If $C2$ s $C3$, then Cn s′ $C3$. Now note that this is the case all the way up to n. Poor, unfortunate Cn cannot scratch anybody in the list already specified. If Axiom 4 must be fulfilled, there must be a new cat to be added to the list; we will call him $C(n + 1)$. Therefore, Cn is not the last cat. Now in the same way that we argued that Cn could not be last, we argue that $C(n + 1)$ cannot be last. Hence there is no last cat. This proves the second alternative:

Theorem 2: There are infinitely many cats in this system.

(There is one small flaw in the proof. See if you can find it. How, if at all, does it affect Theorem 2?)

If we now turn our attention to the first question, that is, "How many cats does a given cat scratch?" we can see that this is a question that our theory cannot answer. (Why?) It is an undecidable issue within the system. We would have to add some more axioms or resort to some other type of device if we wished to make this answerable. One possible theorem is given by the following.

Theorem 3: If each cat scratches at most one cat, then given any cat C there exists exactly one cat B such that C s B.

This theorem provides such an obvious result that it is hardly worth mentioning. The aesthetics of mathematics would rule this theorem as less valuable, for example, than the previous two theorems.

Let us now assume that Axioms 1, 2, 4, and 5 hold and that Axiom 3 is replaced by Axiom 3*: If two cats *A* and *B* are scratched by the same cat, say *C*, then *A* and *B* are the same cat.

Which, if any, of the above theorems remain valid in this system? How many cats does a given cat scratch? Assume now that Axioms 1, 2, 3*, and 4 hold. Show that Axiom 5 is either deducible as a theorem or not deducible as a theorem.

You can already see a difficulty of experiential axiomatics. There are some rather interesting things about this particular set of axioms and the number of cats that are available. Using deductive logic by the axiomatic method, one can show that there must be an infinite number of cats. That is, if any cat at all satisfies these axioms, then there must be an infinite number of them that also satisfy the axioms. This is just one of the difficulties that experiential axiomatics can lead to. One might know, for example, that there are not an infinite number of cats in the world and this fact would seem to imply that there is something wrong with the axiom system. On what are the axioms based? Probably they are based on information obtained by limited experience. It then becomes apparent that the formal properties of these cats and the actual things in the world do not necessarily match.

The first reaction in a case like that above might be to attack the form; but the form is correct. The problem is that the content does not fit the form. If we substitute symbols without content for content-laden terms like *cats* and *scratches*, we will find that the system works flawlessly.

1. There exists at least one object *X*.
2. There exists an *X* such that no other *X* bears the relation *R* to it.
3. If two objects *X* and *Y* bear the relation *R* to the same object *Z*, then the objects *X* and *Y* are the same.
4. Each *X* bears the relation *R* to at least one other *X*.
5. No *X* bears the relation *R* to itself.

Theorem: There are infinitely many *X*'s.

We determined that the problem was one of content not fitting form. We have established the form as being correct, so now the question is, "Can we find content that will fit the form?" Yes, we can. In the next example we leave the form unchanged, but insert new content.

1. There exists a year A.D.
2. There exists a year A.D. which is not preceded by any other year A.D.
3. If two years A.D. precede the same year A.D., the former two years are the same year A.D.
4. Each year A.D. precedes at least one other year A.D.
5. No year A.D. precedes itself.

We have looked at several examples of experiential and formal axiomatics. In point of fact, we have looked at examples of experiential and formal mathematics, and it is on this general note that we will end this section.

There is a difference, a very distinct difference, between concrete and formal mathematics. As defined by Halmos (see reading), formal axiomatics is an example of mathology. Formal mathematics states that any entities X and relation R that happen to satisfy axioms must also satisfy the theorems that you are able to produce. But nothing is said about X and R as such. That is, there is nothing stated about what X and R might happen to represent. Further, since these axioms are propositional forms, truth, factual truth, has absolutely nothing to do with it. Formal mathematics is not concerned with the factual truth nor with the content of its axioms.

On the other hand, there are areas even within mathematics that can be very concerned with truth and content. Experiential axiomatics, or the concrete interpretation of formal axiomatics, assert specific relationships between specific classes of objects.

Summary and a Look Ahead

Even though we have established the meaning and use of the axiomatic method, we must explore a few fine points we have glossed over, repeat some other points, and start thinking about the subject of our next chapter. We must warn you, though, that there will remain many points we will continue to gloss over because of their complexity.

We have neglected to tell you that at the heart of the axiomatic method is the statement, or statement form, "if________, then________." In other words, we never assert that something is true. We assert that *if* something is true, *then* other things will also be true. We are working here with a very particular type of proposition or propositional form, and the framework within which we are working is called logic.

Logic, you will recall, was roughly equated with the grammar of a language. It provided the rules by which one could structure his thoughts into a pattern. Thus, logic is a tool that permits us to draw *valid* conclusions from whatever assumptions are included in our system of axioms, whenever such conclusions exist.

Other than the logic used to work with axiomatic systems, there are certain features that all systems have in common. These features are undefined terms, defined terms, unproven propositions, and proven propositions.

Undefined terms seem to be a necessary evil in axiomatic systems. When you try to define a term or word, you must define it with other words. These secondary words must then be defined with a third set of words. Specifically, one problem is that there is a finite number of words. Within the set of all words there is an even smaller set that can be used to define a given word. This means that eventually, in the process of definition, we will define the word in question by itself.

Consider, as an example, the situation of a foreigner who, even though he has a reasonable command of the English language, encounters the phrase "flat tire" mentioned earlier. In this case the word "tire" is easily defined. However, let us see what difficulties ensue with trying to define "flat" and having it fit into the context. The meanings listed in various dictionaries include: dull, tame, insipid, level, horizontal, and absolute. It is easy to see that if "flat tire" is not understood, can "dull tire," "tame tire," "level tire," or "absolute tire" be any more easily understood?

The second feature that we mentioned was the undefined term. After we establish our set of undefined terms, we are completely free to use them to define other terms.

Borrowing from the Euclidean example, if we accept the term point as being undefined, we can define the term line in reference to point. However, in most instances line is also left undefined. On the other hand, if line is undefined we can go on to define point. We can, in fact, define anything we like in terms of our undefined terms.

The third feature was the unproven proposition. A proposition, you remember in primitive interpretation, is a declarative sentence that is either true or false. You will also remember that an axiom is a statement that is assumed to be true. These unproven propositions are the axioms of the system. To prove a proposition you must employ other propositions which must be proven. The propositions used to prove the propositions proving the proposition must be proven, et cetera, et cetera. To avoid the problem of proving all the propositions, you select the smallest number of propositions necessary to the specific system and assume their truth.

It might be worth mentioning that at this point it becomes necessary to determine whether the axiomatic system in question is experiential or formal. In many cases, those systems that appear at face value to be experiential are in fact formal. This situation arises in spite of the fact that the author of the system intended the system to be experiential. If it is formal, then there will not be propositions but propositional forms. If the system is experiential there will be propositions with content.

The whole point of an axiomatic system is to establish grounds upon which a proposition or propositions can be proven. The proven proposition or theorem is the final feature held in common by all axiomatic systems and is the product of these systems. A theorem is the conclusion of an argument. If valid, it is capable of being deductively proven from the premises or axioms. If a statement is not proved, it is not a theorem but a conjecture.

Having established the general structure of axiomatic systems, we can move on to some of the finer points. There are three points by which you can judge whether or not an axiomatic system is good. If one of these is violated, it can result directly or indirectly in invalid theorems.

In an axiomatic system, consistency is vitally necessary. The axioms must not contradict each other, and you must not be able to prove theorems which contradict the axioms or each other.

Secondly, there is the question of whether or not you can actually prove something from the set of axioms. If the axioms cannot be manipulated by logical means into a hitherto unassumed and unstated proposition, then it is not an axiomatic system, but a group of random assertions.

The last point to check is whether or not the axioms refer to something. This is not to say that all axiomatic systems must be material systems, but that all axiomatic systems must contain the possibility that something could conform to their stipulations.

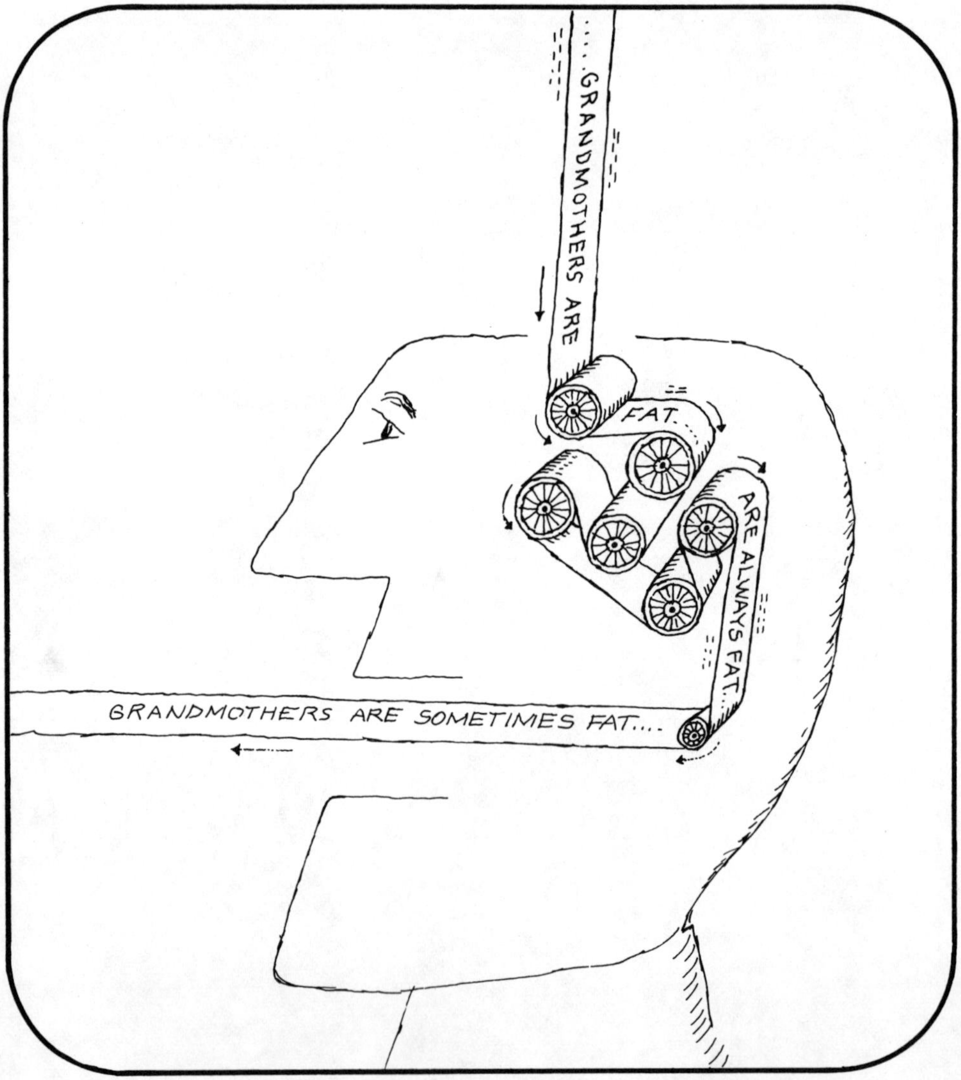
....GRANDMOTHERS ARE
FAT.
ARE ALWAYS FAT....
GRANDMOTHERS ARE SOMETIMES FAT....

Chapter Four

Logics

"I know what you're thinking about," said Tweedledum, "but it isn't so, nohow."

"Contrariwise," continued Tweedledee, "if it was so, it might be, and if it were so, it might be; and if it were so, it would be, but as it isn't, it ain't. That's logic."

Through the Looking Glass

Lewis Carroll

In the last chapter the idea of an axiom system was explored. We saw that an axiom system consisted of defined and undefined terms, axioms, and theorems. We also saw that theorems were deduced from the axioms and previously deduced theorems, if any. However, how does one know when a theorem has been correctly deduced? This chapter attempts to speak to some of the mechanics of that problem.

What is a Logic?

Recall that we have claimed that mathematics possesses a threefold nature. Hence, mathematics appears to be a language, a system of thought, and a set of organized experiences. You should realize by this time that this analysis might just as well apply to your own academic interest. Thus mathematics is only one model. However, let us now focus on the view of mathematics as thought systems.

First, one should not think of mathematics as only one thought system. Quite the contrary, mathematics contains many thought systems. These thought systems are in many cases quite similar. However, in some cases they seem to contradict each other. For example, Euclidean geometry and non-Euclidean geometry seem contradictory in certain ways. On the other hand, as thought systems, mathematical systems have many characteristics in common. Each has propositions, and along with the propositions there are ways to obtain new propositions from already obtained propositions. Moreover, there is a way to tell when some given proposition is "correctly" obtained from some of the others. In other words, mathematics at this level is composed of theories. Theories are thought systems with logics. Notice that it is not proper to speak of just one logic to be applied to all thought systems, for in the most general case each thought system will have its own logic. This even occurs in mathematical thought systems, though not as frequently. The remarkable thing about mathematical thought systems is the similarity of the logics they employ.

There is a strongly defended position that mathematics and logic are the same thing.* However, this position seems to depend on defining these two areas in a

* See W. V. O. Quine, *Mathematical Logic*. Harper Torchbook, 1962. Also see the next chapter.

manner that forces their equivalence. This is generally undesirable, particularly when one is first beginning to think about the nature of mathematics. Part of the confusion in terms arises from a dependence of the one on the other. That is to say, mathematics would find itself having difficulties without the tools logic provides. On the other hand, logic, particularly modern logic, cannot advance without the methods of mathematics.

In one sense, logic may be defined as a set of rules governing the correct production of one proposition from another proposition. Here mathematics and other subject matter areas are quite dependent on some logic. However, logic may be considered as a thought system or better yet, thought systems. We can have thought systems of logics which come equipped with their own logics, etc. A similar position seemingly is espoused by Quine, for he distinguishes two cases of logic:

> . . . logic in the narrower sense comprises those truths which contain only the so-called logical vocabulary essentially, while logic in the broader sense includes both logic in the narrower sense and discourse about it. Discourse of the latter kind is classifiable in a large part at least, under the head of *formal* syntax. . . .

In the brief encounter with logic that you will have in this chapter, you will be introduced to several ways of studying logic. One of the logics that you will be viewing will be that which is nearest the logic used by mathematicians in doing mathematics. Do not be misled into believing that these logic systems are the only kinds that are used in mathematics. For one of the most dramatic examples of this fact see the work of L. E. J. Brouwer and A. Hayting in the area of mathematical intuitionism. (See *both* for an exposition of intuitionistic logic.)

It would be of great interest to explore the logics that come from the nonwestern world. A good example might be provided by India. It is a failure of western civilization today that it cannot understand the nonwest. The primary reason for this is not always linguistic, important as that may be; it is simply because the western mind expects the nonwestern mind to use the same logic systems and the nonwest expects its own logical forms to be known and understood by the western mind. It is a fact that, though the systems are sometimes parallel, they do not intersect where one might expect them to intersect, if at all.

We shall, however, center our study of logic systems on those that are called Aristotelian. More explicitly, we shall say that a logic system is *Aristotelian* if we assume the following to be the case:

(a) For every proposition P, P is the same as P;
(b) For every proposition P, we have either P or not-P (excluded middle);
(c) For every proposition P, we have at most one of P or not-P.

Notice that here and elsewhere we have made no real attempt to "define" what we mean to be a proposition. The preliminary idea of a proposition as was presented in the introductory chapter is good only up to a point. It should be clear, however, that whatever a proposition is in the most general setting, here it is the basic unit of whatever logic system we are considering. Thus, as in the previous chapter, we will take the term "proposition" as being undefined.

Logic, we have said, is the grammar of thought. The study of logic centers about distinguishing between correct and incorrect reasoning. There are several things that should be said about logic at this time. First is a statement by Bertrand Russell to the effect that "Logic, broadly speaking, is distinguished by the fact that its propositions can be put into a form in which they apply to anything whatever." This is empirically supported in many simple cases by the fact that when reasoning correctly from "true" premises you arrive at "true" conclusions, and when reasoning correctly from "false" premises you arrive at conclusions which may be either "true" or "false." If it should happen that you begin with true premises and the conclusion is false, then an Aristotelian-like logic would indicate that the reasoning was faulty. Again, form is the key.

What we want to do here is to formalize logic and to treat it as a mathematical system and, in the process, achieve Russell's goal of applying to anything. We have stated that a proposition is many times taken as any statement which *may* be asserted to be true or false. Examples of this would be sentences such as, "Spring is a season," "Eight is a prime number," or "The one-millionth digit in the decimal expansions of pi is seven."

Basic Assumptions Concerning Logic

All of the above examples have content. To approach Russell's ideal, we will have to move away from content. As in the case of non-Euclidean geometries, this becomes a first step towards a non-Aristotelian logic. In fact, we must completely divorce ourselves from content, and look at propositional forms rather than propositions. Hence, instead of taking "propositions" as our basic undefined term, we shall use the idea of "propositional form!" In addition let us take the word "true" as being undefined. We then make the following definition:

Definition: A propositional form is said to be "false" if and only if it is not true.

Concerning the usage of the words "propositional form," "true," and "false" we assume the following axioms.

Axiom 1: For every propositional form P, P is identical with itself.

Axiom 2: For every propositional form P, P can have the value true or false.

Axiom 3: For every propositional form P, P can have at most one of the values true or false.

Atomic and Molecular Forms

To motivate any kind of formalism (armed with the above axioms) we might be able to achieve, recall that at its lowest levels logic is concerned with simple declarative sentences, and with ways of compounding them into other statements. In ordinary English we have several common connectives, some of which are the words "not," "and," "or," and "if-then." "Not" will be called a *singulary* connective because it is used with a single proposition. "And," "or," and "if-then" are called *binary* connectives because they are used with two propositions.

The connective "not" will be symbolized by a very simple symbol, a capital N. The connective "and" will be denoted by a wedge, $\wedge$, and the connective "or" by a "vee," $\vee$. The connective "if-then" on occasion has been replaced by the simple word "implies." However, some unrestricted use of this convention has led to confusion. Hence, we shall avoid the use of "implies." We use instead the combination "only if" as equivalent to "if-then." In any event, we shall use "$\rightarrow$" for the symbolization of "if-then" or "only if."

The first type of propositional form is that which we shall call an atom; the second type will be called a molecule. An *atom*, or *atomic propositional form*, is one which has no connectives. A *molecular propositional form* is then one which does have connectives. If we were to write something like "Spring is a season," and represent it by p, we would have an atomic propositional form.

The following are ways to build the simplest types of molecular propositional forms. To use the connective "not," you precede the form you are denying with N, written Np. The connective "and" is placed between the forms being connected. So, if p and q are forms, the form "p and q" is denoted by $p \wedge q$. In similar fashion, for two forms p and q, the form "p or q" is denoted by $p \vee q$; while the form "p only if q" is presented as $p \rightarrow q$.

Tables of Form Values

In many conversations the use of these connectives leads to ambiguous interpretations. In the case of denial this is precipitated by the irregularity of the denial process in the English language. Sometimes the word "not" is attached to the main verb of the sentence. Thus, the denial of "Grandmothers are fat" is "Grandmothers are not fat." If the verb is modified by words such as "sometimes" or "always" the denial is formed by using "never" or "not always." Thus, "Grandmothers are sometimes fat" is denied by "Grandmothers are never fat." (As an exercise, form the denials of "Grandmothers are always fat" and "Grandmothers are never fat.")

However, one of the major roles of logic is standardization by the introduction of symbols. Thus, for the denial of

1. Grandmothers are fat,
2. Grandmothers are sometimes fat,
3. Grandmothers are always fat,
4. Grandmothers are never fat,

we write

N Grandmothers are fat,

or

N (Grandmothers are fat),

etc. The parentheses indicate grouping. Consequently, the denial of a statement is true if and only if the original statement is false. Similarly the denial of a statement is false if and only if the original statement is true. This may be presented in Table 4-1.

Table 4-1

Symbol	Form p	Denial of form Np
Value	T	F
	F	T

Let us now look at the problem of combining forms in pairs. In this instance we have two forms to be put together to yield a new form. We would like the truth value of the combined form to be determined by the values of the component forms. In the case of "and," ordinary usage suggests that the combined form is true when and only when both components are true, and false in each of the other cases of which there are three. This is summarized in Table 4-2 for two propositional forms p and q:

Table 4-2

Symbol	1st Form p	2nd Form q	Conjunction $p \wedge q$
Value	T	T	T
	T	F	F
	F	T	F
	F	F	F

Our next combined form involves the use of "or." There are two common meanings attached to this word. In one case the combined form is true if at least one of the alternatives is true, and false when both of the alternatives are false. This case is called the inclusive case and shall be the one we shall focus on. Its values are shown in Table 4-3:

Table 4-3

	1st Form	2nd Form	Alternation
Symbol	p	q	$p \vee q$
Value	T	T	T
	T	F	T
	F	T	T
	F	F	F

The so-called *exclusive* case is the same except for the top-value row. This case insists that the alternation is true if *exactly* one of the alternatives is true. (Exercise: denote the exclusive "or" by $\underline{\vee}$ and make a table of truth values covering this case.) A word of caution is needed here. In ordinary conversation the words "and" and "or" are used to connect all sorts of things other than statements. Their symbolic usage as given above refers only to their use as *sentential* connectives.

Let us now consider the connective "if-then," or "only if." Sentences of this type are called *conditionals* and their formation is somewhat irregular in English depending upon the style of the writer. For example, suppose we are given the following simple sentences:

p: Cows are animals;

q: The ocean is salty.

The following are equivalent expressions of the same conditional situation:

1. $p \rightarrow q$;
2. Cows are animals only if the ocean is salty;

3. If cows are animals, then the ocean is salty;
4. That cows are animals is a sufficient condition for the ocean to be salty;
5. A necessary condition for cows to be animals is that the ocean is salty.

When expressed in the form $p \rightarrow q$, p is usually called the *antecedent* and q the *consequent*. A number of properties of the conditional are apparent from ordinary usage. Any conditional with true antecedent and true consequent is a true statement; most important, if a conditional has true antecedent and false consequent, then the conditional is false. These truth values are summarized in Table 4-4.

Table 4-4

	Antecedent Form	Consequent Form	Conditional Form
Symbol	p	q	$p \rightarrow q$
Value	T	T	T
	T	F	F
	F	T	—
	F	F	—

Thus the conditional "if cows are animals, then the ocean is salty" is false, or "disproved," if it can be shown that cows are animals and that the ocean is not salty. However, what can we say of the last two cases illustrated in the truth table?

If cows are not animals, can the ocean still be salty? Does the form of the conditional give us any information? Let us consider another example: "If Kilroy was here, then the mark is his." Here the situation is somewhat clearer. If Kilroy was not here, then both consequent values *can* be made acceptable in ordinary usage even though the situation is somewhat ambivalent. Hence to remove ambiguity we shall give a compromise to approximate ordinary usage. To show this, we have Table 4-5:

Table 4-5

	Antecedent Form	Consequent Form	Conditional Form
Symbol	p	q	$p \to q$
Value	T	T	T
	T	F	F
	F	T	T
	F	F	T

As an exercise write a paper commenting upon the following idea: "Mathematics and logic are approximations which serve as compromises between the way we would like things to be and the way things are or appear to be."

A Variation

Let p and q be propositional forms. In the molecular propositional form $p \wedge q$, both p and q must be true for $p \wedge q$ to be true. That is, taken by themselves, p must be true, and q must be true. For example, if one were to assert "It is cold and it is wet," and in fact, it was hot, then the statement as a whole would be false.

Now observe the following cases when p and q represent true situations and hence Np and Nq represent false situations:

$$p \wedge q \qquad p \wedge Nq$$

$$Np \wedge q. \qquad Np \wedge Nq.$$

Clearly, of the four only the first represents a truth.

"Or" statements as we have seen above present a different problem. If either alternative is true, then the entire statement is true. Hence, taking p and q as above we have the possibility of the following alternations:

$$p \vee q \qquad p \vee Nq$$

$$Np \vee q \qquad Np \vee Nq.$$

Of these only $Np \vee Nq$ is false while the other three forms are true.

We shall leave it to the reader to construct a similar analysis for "only if."

Truth Tables for Compositions

As an added benefit from the last section we see one of the ways two connectives can be used in the same sentence. We shall see other ways here and show how to construct truth-value tables for them. Consider the following sentence: "It is not the case that Kilroy was here or the mark is his." This may be partially symbolized by

N(Kilroy was here) $\vee$ (the mark is his).

If we denote "Kilroy was here" by p and "the mark is his" by q, we have a complete symbolization given by $Np \vee q$. Note that the N applies only to p and not to q. Let us look at Table 4-6:

Table 4-6

	1st Form		2nd Form	Alternation
Symbol	Np	p	q	$Np \vee q$
Value	F	T	T	T
	F	T	F	F
	T	F	T	T
	T	F	F	T
	*		*	

Since the forms in the columns marked with the (*) are the alternatives for the last column these are the values paired, not the values under p. This is a preliminary result used to produce Np. As an exercise discuss any possible similarity existing between this table and the one for the conditional form, $p \rightarrow q$.

Now let us compare the above with N((Kilroy was here)$\wedge$ N(the mark is his)). (Make sure you understand the use of parentheses as separators.) Our table is thus displayed as

Table 4-7

	1st Form	2nd Form		Conjunction	Denial of Conjunction
Symbol	p	q	Nq	$p \wedge Nq$	$N(p \wedge Nq)$
Value	T	T	F	F	T
	T	F	T	T	F
	F	T	F	F	T
	F	F	T	F	T
	*		*	**	

As a more complex example consider $(p \rightarrow q) \wedge p$. We shall display the values in stages beginning with Table 4-8:

Table 4-8

1st Form $p \rightarrow q$	2nd Form p	Conjunction $(p \rightarrow q) \wedge p$
—	T	—
—	T	—
—	F	—
—	F	—

We note that the first form is itself decomposible into simpler forms p, q, for which we have already assigned values to p. Hence, adding a column for q, we have Table 4-9:

Table 4-9

Help		1st Form	2nd Form	Conjunction
p	q	$p \rightarrow q$	p	$(p \rightarrow q) \wedge p$
T	T	T	T	T
T	F	F	T	F
F	T	T	F	F
F	F	T	F	F

Consider now the following form $(p \rightarrow q) \wedge p) \rightarrow q$. Table 4-9 can be used to construct Table 4-10:

Table 4-10

Help		1st Form	2nd Form	Conditional
p	q	$(p \rightarrow q) \wedge p$	q	$((p \rightarrow q) \wedge p) \rightarrow q$
T	T	T	T	T
T	F	F	F	T
F	T	F	T	T
F	F	F	F	T

Note that the main column of Table 4-10 has the value T for all possible values of the constituent forms. Any form having these properties is called a *tautology*. The following are examples of tautologies:

1. $((p \rightarrow q) \wedge Nq) \rightarrow Np$
2. $((p \vee q \wedge Np) \rightarrow q$
3. $((p \vee q) \wedge Nq) \rightarrow p$

4. $p \to (p \vee q)$
5. $q \to (p \vee q)$
6. $(p \wedge q) \to p$
7. $(p \wedge q) \to q$
8. $((p \to q) \wedge (q \to r)) \to (p \to r)$

We leave it to the reader to construct the appropriate tables to demonstrate that each is a *tautology*. If all values are false then we have what is called an *absurdity*. Show that the form $p \wedge Np$ is an absurdity, notice that (strange as it may seem) $p \to Np$ is not an absurdity but show that $(p \to Np) \wedge (Np \to p)$ is. It is interesting to note that $(p \to Np) \vee (Np \to p)$ is a tautology. Explain completely what is happening here.

Tautologies are particularly important to us because they provide one way to decide the validity of an *argument*. For our purpose, an argument is composed of two sets of statements, one set called premises and the other set called conclusions.

Let $p = \{p_1, p_2, \ldots, p_n\}$ be the set of premises for an argument and $c = \{c_1, c_2, \ldots, c_k\}$ the set of conclusions. Then the argument is said to be *valid* if and only if

$$(p_1 \wedge p_2 \wedge \cdots \wedge p_n) \to (c_1 \wedge c_2 \wedge \cdots \wedge c_k)$$

is a tautology. This can be denoted by $p_1, p_2, \cdots, p_n \vdash c_1, c_2, \cdots c_k$. If an argument is not valid, it is said to be *invalid*. *Note*: In most of the cases in which we are interested one possible conclusion will be selected. Thus only an expression such as $p_1 \wedge p_2 \cdots \wedge p_n \to c_i$ will be examined. It would also seem in light of certain possible application that perhaps when one has a number of possible conclusions that one might wish to test for tautology on expression of the form

$$p_1 \wedge \cdots \wedge p_n \to c_1 \vee (c_1 \wedge c_2) \vee (c_1 \wedge c_2 \wedge c_3) \vee \cdots$$

Now as an example suppose that

p_1 is $p \vee q$

p_2 is $p \to r$

p_3 is $q \to r$

Further suppose that c_1 is $r \vee s$. We shall show that $(p_1 \wedge p_2 \wedge p_3) \to c_1$ is a tautology. We must do this, of course, by the construction of a truth table. First we notice that we have four distinct propositional forms as atoms: p, q, r, s. Hence we must have more possible combination of T and F than before. (See Table 4-11.)

Table 4-11

p	q	r	s	$p \vee q$	$p \to r$	$q \to s$	$P_1 \wedge P_2 \wedge P_3$	$p \to c$	$r \vee s$
T	T	T	T	T	T	T	T	T	T
T	T	T	F	T	T	F	F	T	T
T	T	F	T	T	F	T	F	T	T
T	T	F	F	T	F	F	F	T	F
T	F	T	T	T	T	T	T	T	T
T	F	T	F	T	T	T	T	T	T
T	F	F	T	T	F	T	F	T	T
T	F	F	F	T	F	T	F	T	F
F	T	T	T	T	T	T	T	T	T
F	T	T	F	T	T	F	F	T	T
F	T	F	T	T	T	T	T	T	T
F	T	F	F	T	T	F	F	T	F
F	F	T	T	F	T	T	F	T	T
F	F	T	F	F	T	T	F	T	T
F	F	F	T	F	T	T	F	T	T
F	F	F	F	F	T	T	F	T	F
				*	*	*	***		

As another example consider (The U.S. will withdraw $\vee$ (Red China will invade $\wedge$ South Korea will retaliate)) $\to$ ((The U.S. will withdraw $\vee$ Red China will invade) $\wedge$ (The U.S. will withdraw $\vee$ South Korea will retaliate)). In other words, $(p_1 \vee (p_2 \wedge p_3)) \to ((p_1 \vee p_2) \wedge (p_1 \vee p_3))$. We shall show that this is a tautology.

Table 4-12

p_1	p_2	p_3	$p_2 \wedge p_3$	$p_1 \vee (p_2 \wedge p_3)$	$(p_1 \vee p_2)$	$p_1 \vee p_3$	$\wedge$	$\rightarrow$
T	T	T	T	T	T	T	T	T
T	T	F	F	T	T	T	T	T
T	F	T	F	T	T	T	T	T
T	F	F	F	T	T	T	T	T
F	T	T	T	T	T	T	T	T
F	T	F	F	F	T	F	F	T
F	F	T	F	F	F	T	F	T
F	F	F	F	F	F	F	F	T
				*			*	

Note also that $((p_1 \vee p_2) \wedge (p_1 \vee p_3)) \rightarrow (p_1 \wedge (p_2 \wedge p_3))$ is a tautology. As an exercise demonstrate this.

Some Mathematics is Logic

The above method of dealing with problems of logic, though fairly straightforward, can be somewhat tedious when dealing with a large number of different propositional forms. In this section we shall introduce you to an approach to logic that is more mathematical in nature. In fact it imitates formal mathematics to a very strong degree. In the exercises we shall provide, you will see in a very simple form all the elements of mathematics in action. Each of the problems shall be miniature mathematical systems complete with the major ingredients: undefined terms, axioms, and theorems. In some cases you will be asked to prove a stated theorem, while in others you will be asked to supply some theorems to which the axioms can lead.

As a means by which these objectives may be accomplished, we shall list below a set of rules called "rules of inference." These rules will permit you to operate on the items of a problem so that proofs may be constructed. If you examine the rules carefully, you will note that for the most part they are restatements of the items listed above as tautologies.

The rules are as follows:

1. From the form $p \wedge q$ it is correct to conclude both the form p and the form q.
2. From the form p and the form q it is correct to conclude the form $p \wedge q$.
3. From the form p it is correct to conclude $p \vee q$. Similarly from q one obtains $p \vee q$.
4. From the form $p \rightarrow q$ it is correct to obtain the form $Nq \rightarrow Np$.
5. From the form $p \rightarrow q$ and the form p it is correct to obtain the form q.
6. From the form $p \rightarrow q$ and the form Nq it is correct to obtain the form Np.
7. From the form $p \vee q$ and the form Np it is correct to obtain the form q. Similarly, from $p \vee q$ and Nq one obtains p.
8. From the form p it is correct to obtain the form NNp. From the form NNp it is correct to obtain the form p.
9. From the forms $p \rightarrow q$ and $p \rightarrow r$ it is correct to obtain $p \rightarrow r$.

Note that it is incorrect to obtain the form p from the form $p \rightarrow q$ and the form q. This situation can easily be seen from the construction of a truth table for $((p \rightarrow q) \wedge q) \rightarrow p$. The latter form is not a tautology; hence, it is not acceptable.

We shall tacitly assume that any of these rules of inference can be applied as often as needed and that once an item has been derived by a rule of inference it may be used to obtain some other item.

Suppose now that $r \rightarrow Nt$, $s \rightarrow r$, and s are given forms. Let us see what the rules of inference permit us to conclude. First arrange these given items in a column. We shall always use the following form:

$$\begin{array}{l} r \to Nt \\ s \to r \\ \underline{s \qquad\quad} \end{array}$$

As a general rule for trying to find something with which to start, we usually examine all possible pairs of given items until we find a situation that fits one of the rules. Thus you may note in the above example the pair

$$\begin{array}{l} s \to r \\ r \to Nt \\ \underline{s \qquad\quad} \end{array}$$

permits us to conclude $s \to Nt$ by Rule 9. We denote this situation by

$$\begin{array}{ll} s \to r & \\ \underline{r \to Nt} & \\ s \to Nt & \text{by Rule 9.} \end{array}$$

Now using the idea stated above that once an item has been obtained by the use of some rule we may use it as though it was a given item, we note

$$\begin{array}{ll} s \to Nt & \\ \underline{s \qquad\quad} & \\ Nt & \text{by Rule 5.} \end{array}$$

Thus we see, using all three of the given items, leads to the theorem Nt. The *proof* consists of the steps used to obtain the result. Now observe that this sequence of steps is not the only ones that could have been used to obtain Nt. The following sequence gives also Nt:

$$\begin{array}{ll} s \to r & \\ \underline{s \qquad\quad} & \\ r & \text{by Rule 5.} \\ r \to Nt & \\ \underline{r \qquad\quad} & \\ Nt & \text{by Rule 5.} \end{array}$$

If the connective $\wedge$ occurs, a slightly different approach must be taken. For example, show that from the forms

$$Nr \wedge t$$

$$s \rightarrow r$$

we may obtain Ns. Here we note as a first step that from $Nr \wedge t$ we may obtain both Nr and t by Rule 1. Thus Rule 6 shows

$$\begin{array}{l} s \rightarrow r \\ Nr \\ \hline Ns \end{array}$$

As another example, prove that the items above the line lead to the item below the line in

$$\begin{array}{l} p \vee q \\ Nq \\ p \rightarrow s \\ \hline s \quad t \end{array}$$

Proof:

$$\begin{array}{l} p \vee q \\ Nq \\ \hline p \end{array} \qquad \text{by Rule 7.}$$

Then

$$\begin{array}{l} p \rightarrow s \\ p \\ \hline s \end{array} \qquad \text{by Rule 5.}$$

However Rule 3 tells us that from s we may obtain $s \vee t$ where t could be any propositional form.

Exercises

In each of the following if possible show that the indicated results can be obtained. If no conclusion is indicated, produce one using as many of the given items as possible.

1. $r \to Nt$
   $s \to r$
   s
   ———
   $Nt \wedge r$

2. $Nh \to Nj$
   Nh
   $Nj \to s$
   ———
   ?

3. $a \to (b \wedge d)$
   $(b \wedge d) \to c$
   a
   ———
   c

4. $Ng \to e$
   $e \to k$
   Ng
   $k \to Nl$
   $Nl \to m$
   $m \to b$
   ———
   b

5. $s \to t$
   s
   ———
   NNt

6. Na
   $Na \to NNb$
   ———
   b

7. $r \to NN(p \vee q)$
   r
   ———
   $p \vee q$

8. $j \to (k \land m)$
   j
   $\underline{(k \land m) \to NNq}$
   q

9. $Np \to q$
   $\underline{Nq}$
   ?

10. $q \to Nr$
    $\underline{NNr}$
    ?

11. $g \to h$
    $Ng \to NNf$
    $\underline{Nh}$
    f

12. Nb
    $a \to b$
    $\underline{Na \to c}$
    c

13. If $x = y$, then $y = z$.
    If $y = z$, then $y = w$.
    If $y = w$, then $y = 1$.
    $\underline{y \neq 1.}$
    ?

14. $c \to a$
    c
    $\underline{c \to b}$
    $a \land b$

A Three-Valued Logic

In this section we shall investigate what can happen if we drop the restriction that propositional forms can have only one of two values, T, or F. Let us suppose that a form can be assigned one of three values, T, F, or ?. Our task is now to decide what alteration will take place in the traditional combined forms. To accomplish this analysis, we shall use a slightly modified truth table.

First, let us consider the form $p \wedge q$ and say that $p \wedge q$ has the value T if and only if p is T and q is T. We demonstrate this in Table 4-13:

Table 4-13

		q		
	$\wedge$	T	F	?
p	T	T	—	—
	F	—	—	—
	?	—	—	—

Now we are faced with the problem of how to fill in the eight blanks. Since each cell can be filled in in exactly two ways (T has been eliminated as a possible filler), there are $2^8 = 256$ ways of filling with F and ?. Usage, however, suggests that some ways might be better than others. One possible way is to assume the following:

1. ? is falser than T;
2. F is falser than ?;
3. $p \wedge q$ is exactly as true as the falser of p or q.

Thus we have Table 4-14:

Table 4-14

		q		
	∧	T	F	?
p	T	T	F	?
	F	F	F	F
	?	?	F	?

Notice that the T–F part of the table is identical to the classical interpretation.

Suppose that $p \wedge q$ is T if and only if p is T and q is T. Suppose also that ? means undecidable. Complete Table 4-13. Assume that whenever p or q are ?, then $p \wedge q$ is ?.

Table 4-15

		q		
	∧	T	F	?
p	T	T	—	—
	F	—	—	—
	?	—	—	—

The introduction of a third possible value also complicates the issue for denial. Let us assume the following:

1. If p is T, then Np fails to be T.
2. If p is F, then Np fails to be F.

The following is a possible table

Table 4-16

p	Np
T	?
?	F
F	T

Notice that our old friend "double negation," that is $p = NNp$, fails in this logic. [$p = q$ means $(p \to q) \wedge (q \to p)$ is a tautology.]

Exercises

1. Show that in this logic $p = NNNp$. (*Hint:* In a two-valued logic the proof that $p = NNp$ reduces to showing $(p \to NNp) \wedge (NNp \to p)$ is a tautology. This is accomplished by examination of column (**).

$(p$	$\to$	N	N	$p)$	$\wedge$	$(N$	N	p	$\to$	$p)$
T	T	T	F	T	T	T	F	T	T	T
F	T	F	T	F	T	F	T	F	T	F
	*				**				*	

Thus to solve the above problem we apparently must show that $(p \to NNNp) \wedge (NNNp \to p)$ is a tautology in a three-valued logic. We have a way of computing ($\wedge$); in fact we have two ways. However, we don't seem to have defined a truth-value table for ($\rightarrow$) in a three-valued logic. How would *you* do it? There is another approach for the faint-hearted. Show that the truth tables for p and $NNNp$ are the same by computing the tables for p, Np, NNp, $NNNp$ successively.

2. Find a real-life example where this might be a useful logic.

Since in a two-valued logic we like to have the equivalence of p and NNp, the question arises, "Does there exist a three-valued logic in which p and NNp are the same?" The answer is "yes!" Look at Table 4-17:

Table 4-17

p	Np	NNp
T	F	T
?	?	?
F	T	F

To complete our catalogue of connectives introduced in two-valued logic, we must consider the forms $p \vee q$ and $p \rightarrow q$. How do they behave in a three-valued situation? One approach is to define $p \vee q$ to be $N(Np \wedge Nq)$. This relationship is the case in our two-valued logic; *define* it to be the case in a three-valued logic. Let us construct a table for $p \vee q$, using Table 4-14 and Table 4-17.

Table 4-18

p	q	Np	Nq	$Np \wedge Nq$	$N(Np \wedge Nq)$
T	T	F	F	F	T
T	F	F	T	F	T
T	?	F	?	F	T

Table 4-18 (continued)

p	q	Np	Nq	$Np \wedge Nq$	$N(Np \wedge Nq)$
F	T	T	F	F	T
F	F	T	T	T	T
F	?	T	?	?	?
?	T	?	F	F	T
?	F	?	T	?	?
?	?	?	?	?	?

Thus, we have Table 4-19 for $p \vee q$:

Table 4-19

	$\vee$	q: T	T	?
p	T	T	T	T
	F	T	F	?
	?	T	?	?

3. Define $p \to q$ to be the same as $Np \vee q$ and show that the following table is a possible three-valued table for $p \to q$.

Table 4-20

	$\to$	q: T	F	?
p	T	T	F	?
	F	T	T	T
	?	T	?	?

(*Hint:* use Table 4-19.)

4. Using the above exercise, go back and solve Exercise 1, page 79.
5. For those who have been exposed to probability and/or statistics, discuss a possible connection between these ideas and multi-valued logics.

Chapter Five

The Fabric of Historic Mathematics, Part II

"Would you tell me, please, which way ought I
walk from here?"
"That depends a good deal on where you want to get to,"
said the Cat.
"I don't much care where—," said Alice.
"Then it doesn't matter which way you go,"
said the Cat.

Alice in Wonderland

Lewis Carroll

Zeno and His Paradoxes

Lest one think that all was well with mathematics once formal axiomatics had made its occurrence, be assured that there was more trouble in store for the mathematician. This was particularly true of those who felt the responsibility of helping to police mathematics as well as produce it. Do not think that this policing action is restricted to modern times. Let us again turn to the Greeks who had the very bad habit of asking embarrassing questions of themselves and others.

Among the many questions investigated by the Greeks were ones concerning conceptions of time, motion, space, and in particular the concept of a *straight* line. (In modern mathematics we sometimes refer to this latter concept as a *linear continuum.*) All of these concepts are as important today as they were to the ancient Greeks. The difficulties they observed are constantly being revived. Hence, let's examine the questions raised by the Greeks Zeno and Parmenides (400 B.C.). These men lived about the same time as Plato and their ideas are discussed in both Plato and Aristotle.

The questions, or arguments, of Zeno were four in number but we need only be concerned with three of them. We shall refer to the three, listed below, as Zeno's Paradoxes. For a complete discussion of the idea of paradox see Quine's article in the Readings.

1. Motion is not possible. Suppose motion is possible, then in order to cover a certain distance, say 1 mile, one must travel half that distance or one-half mile, and before that, half the last distance and so on. But this means an infinite process must be performed in a finite length of time. Impossible!
2. If a tortoise has a headstart, then a swift runner, say Achilles, will never overtake it, for by the time Achilles gets to the tortoise's starting point the tortoise will be at a further point, and by the time Achilles arrives at this point the tortoise will be at another point, ad infinitum. Thus the slower tortoise will always be ahead!
3. Consider now an arrow in flight, at a given instant the head is at some particular point in space. Since it is at rest at that particular

point it cannot be in motion then. It is easy to see that at every instant it is at rest at some point in space, and hence is motionless.

Look up the fourth of Zeno's Paradoxes. What specific problem does it relate to? After you complete your study of this chapter suggest a means for dealing with this paradox.

What is Zeno saying? Much as Plato, Zeno tried to distinguish between the world of sense perception, or appearance, on the one hand and the real world on the other. Thus, he is not trying to argue against motion as an observed phenomenon; rather, he assumes that the real world is consistent and thus free of contradiction. We also note that the first two paradoxes can be thought of as expressing opposition to the concept of a line being made of infinite many points. The last paradox opposes finite subdivision of the line. But a line is either finitely divisible or infinitely divisible, isn't it? "No!," said Parminides, "a line is indivisible, a continuum." Thus, we have the beginning of the discrete-continuous and particle-wave controversy (the former in mathematics and the latter in physics).

We note also that the paradox concerning Achilles and the tortoise can also be considered as a question concerning the number of points on two line segments of differing length. The "obvious" answer is, if there are points at all, the longer one will have more points. But wait! Consider the set of natural numbers $N = \{1, 2, \ldots, n, \ldots\}$. Now consider the set $N^2 = \{1, 4, 9, \ldots, n^2, \ldots\}$. These two sets have the same number of members. We can see this by pairing each number of the first set with its square. This property was noted by Galileo in his Dialogues in 1638. Later this same property was used by Cantor to define when two sets have the same *cardinal* number. If we can match the elements of one set with the elements of another set in such a way that all the elements of one set correspond to one and only one element of the other set, the correspondence is said to be *one-to-one*. Thus, if there exists a one-to-one correspondence between two sets, the two sets are said to have the same *cardinal number*. Let us look at the following examples.

1. Let $A = \{a, b, c\}$ and $B = \{1, 2, 3\}$. Clearly, A and B have the same cardinal number; there are many ways of setting up the one-to-one correspondence.
2. Let $N = \{1, 2, \ldots, n, \ldots\}$ and $Q = \{p/q : p$ and q are elements of $N\}$. This example is a little more difficult. Get some help on it!

3. Two line segments of possibly different length can have a one-to-one correspondence between their points. This correspondence is partially shown below.

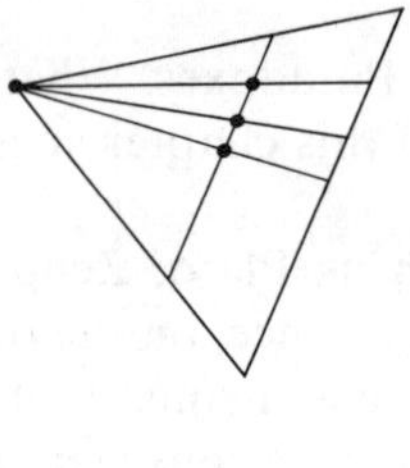

Figure 5-1

The last example shows that the Achilles paradox is no paradox in this regard! Once again, we see that careful definition and clear statement of underlying axioms are somewhat helpful.

On to the Calculus

There is still the matter of distance to be examined; for one of the things Zeno was saying was that the following series of numbers with infinitely many terms must have an infinite sum:

$$1 + \frac{1}{2} + \frac{1}{4} + \cdots + \frac{1}{2^n} + \cdots$$

"Not so!" say moderns, "the sum is 2." But what is meant by "sum" here is quite precise. Here, what we mean is to take the sequence

$$1$$

$$1 + \frac{1}{2} = \frac{3}{2}$$

$$1 + \frac{1}{2} + \frac{1}{4} = \frac{7}{4}$$

$$1 + \frac{1}{2} + \frac{1}{4} + \frac{1}{8} = \frac{15}{8}$$

$$1 + \frac{1}{2} + \frac{1}{4} + \frac{1}{8} + \cdots + \frac{1}{2^n} = 2 - \frac{1}{2^n}.$$

The last statement can be obtained by observing that the series is what is known as a geometric series. Now notice that when n is a very large number, then $\frac{1}{2}^n$ is a very small number. Hence, we notice that at each step, the sum differs from 2 by successively smaller amounts. Hence, in the matter of infinite processes, we are speaking of *becoming* infinite rather than *being* infinite.

The above example serves to point the way to the concept of *limit*. It is one of the basic ideas explored in the modern mathematics of the infinite and of the infinitesimal. These problems, in their elementary form, are studied in a course usually called "The Calculus." The course is misnamed, since the word "calculus" can be taken to mean any method of *calculation*. In fact, in logic one speaks of the "propositional calculus" in talking about how to do calculations involving propositional forms.

We return to the limit concept now to make it more precise. We shall say that a sequence $\{a_1, a_2, \ldots\}$ approaches or has a real member s as a limit if after sufficiently many terms of the sequence, the value of the terms from that point on differ from s by less than any preassigned positive number. This is a strange way to define the concept, but it seems to be just what is needed to tell how to use finite processes to describe and to approximate infinite processes to any desired degree of accuracy.

It is interesting to note that all elementary calculus is built on Zeno's Paradoxes. Not only does calculus involve limit as such, but it involves the concept of motion as in the case of the moving arrow that doesn't move. This branch of calculus is called differential calculus. Let us refer to it as the calculus of differentiation. It is here that we try to make precise the concept of *instantaneous* motion. This is done by considering the action of the arrow, for example, over decreasingly shorter time periods, and using the limit concept to deduce the motion characteristics of the arrow.

The definition of velocity at a given instant is usually given in terms of average velocity over a time interval. The latter is given by the ratio

$$\frac{\text{Displacement}}{\text{Time elapsed}}.$$

This definition says nothing directly about velocity at a particular instant. However, Zeno again gives us a clue. It seems reasonable to define instantaneous velocity in terms of average velocity over short periods of time. That is, average velocity can be used to approximate instantaneous velocity to any degree of accuracy desired by reducing the time intervals.

Thus, suppose an object is dropped from a height of 500 feet. Galileo's law of falling objects says that the displacement d after t seconds is given by the formula

$$d = 16t^2.$$

Let us see what can be said about the instantaneous velocity at $t = 1$ second. We note that the average velocity in feet per second during the time interval $t = 1$ to $t = 1.1$ is given by the formula

$$\frac{\text{Displacement}}{\text{Elapsed time}} = \frac{16 \times (1.1)^2 - 16 \times 1^2}{1.1 - 1}$$

$$= 16 \times 2.1 = 33.6.$$

By similar computations we can construct the following table which lists the average velocity for small time intervals containing $t = 1$.

t	Average velocity between 1 and t
1.1	33.6
1.001	32.016
1.00001	32.00016
.999	31.984
.99999	31.99984

A close examination of this table suggests that the instantaneous velocity should be *defined* to be 32 feet per second at $t = 1$. In further support of this definition, suppose we consider the time interval between time $t = 1$ and time $t = s$ (s for any time). Then the velocity is

$$\frac{\text{Displacement}}{\text{Elapsed time}} = \frac{16s^2 - 16 \times 1^2}{s - 1}.$$

A little algebra shows:

$$\frac{16s^2 - 16 \times 1^2}{s - 1} = \frac{16(s^2 - 1^2)}{s - 1}$$

$$= \frac{16(s + 1)(s - 1)}{s - 1}$$

$$= 16(s + 1).$$

Now, as s becomes 1 the time interval shrinks to zero, and the average velocity becomes 32 feet per second. (The astute reader will notice how cleverly we avoid dividing by zero.) As an exercise to test your own understanding, try to conjecture how the instantaneous velocity would be defined at $t = 3$.

As a second exercise for those of you who have had sufficient mathematical background to understand the basic concept of a function and functional notation, observe that we may generalize the preceding. Suppose that an object moves in such a way that its displacement from some fixed point is $f(t)$ at any time t. Thus, the average velocity between time t and a is given by

$$\frac{\text{Displacement}}{\text{Elapsed time}} = \frac{f(t) - f(a)}{t - a}.$$

Let

$$f(t) = 2t^2 - t$$

and make a reasoned conjecture about how one should define the instantaneous velocity when $t = 0$, and $t = 1$.

Of equal importance in calculus is the problem of the determination of area and related problems. As long as the regions are the well-known ones of triangles, squares, rectangles, and combinations of these basic forms, there is no problem. But even such a simple figure as the circle causes some difficulties. Most ancient people could approximate the area of a circle, using an approximate value, experimentally obtained, for π in $A = \pi r^2$ where r is the radius. Perhaps one of the best procedures for obtaining the area of a circle was by Archimedes (287–212 B.C.) who used successive approximations by inscribed and circumscribed polygons whose exact area he obtained. He then used the limit idea we have been discussing. We shall look only at the "circumscribe" case. We note that the area of a polygon with equal length sides is found by the formula

$$A = \frac{1}{2}r \cdot P$$

where P is the perimeter and r is the perpendicular distance from the center to a side as shown in the hexagon in Figure 5-2.

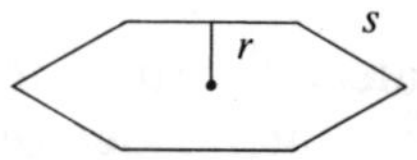

Figure 5-2

Thus, if we know that a circle of radius r can be inscribed in a regular polygon (see Figure 5-3), then increasing the number of sides "indefinitely" produces the area of the

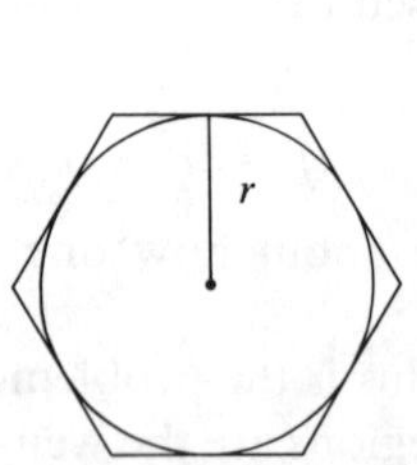

Figure 5-3

circle. This is by using the concept of a limit. (Try to generalize to $y = x^2$. Note the relationship of the above to Zeno's fourth paradox that you were asked to look for.)

On Beyond Non-Euclidean Geometry

Once the fact of the non-Euclidean geometries was acknowledged, it became clear that some drastic steps would have to be taken so that mathematics would never have to suffer such humiliation again. David Hilbert tried to clean up Euclid's geometry with his insistence on establishing a formal axiom system for geometry. His primary aim was to ensure a mathematics free of contradiction. He sought to do this by wedding formal axiomatics and logic. Without a logic, freedom from contradiction is assured only by consistency demonstrations. Demonstrations of consistency by giving a model such as in the Montague and Montgomery article in the readings show that this process tends only to shift the consistency problem from one branch of mathematics to another.

Now while Hilbert was trying to get content out of the way of mathematics, the German logician and mathematician Frege was objecting to some of the primitive versions of formalization. Frege believed that mathematics must rest on the principles of logic. Nothing else was needed. Thus we have a reduction of mathematics to logic. Frege assumed for himself the task of showing how this reduction could be accomplished. To do it he had to give a complete characterization of the principles of logic. He then had to show that all the axioms of mathematics were deducible from the principles of logic. He was reasonably successful at this task until the turn of the century when Bertrand Russell presented Frege with a paradox. Later Russell and A. N. Whitehead tried to resume Frege's work. They attempted to show also that mathematics can be reduced to logic. They tried to overcome the difficulties that Frege had encountered.

However, the logistic program even if successful still leaves open the question of consistency of that logic to which mathematics is reduced. The latter situation occurs because of the axiomatic structure of logic (see Chapter 4). As Hilbert envisioned it, through the formalism suggested by logically analyzing mathematical concepts and processes and representing them in some appropriate symbolism (say symbolic logic), we may be able to avoid contradiction. That is, we may be able to show that a formula and its contradiction can never be obtained from the same fundamental formulae (axioms) and the rules laid down for obtaining derived formulae and manipulating symbols. This, in effect, equates mathematics with mathematical notation. Thus, mathematicians become symbol pushers.

Hilbert was not content just with consistency. He reflected the attitude of the times when he assumed that all problems could be solved. By this he means that any question properly phrased within a given formal system is answerable within that system. In other words, formal systems should be self-consistent and complete. Later in 1931, Kurt Gödel blasted any hope of accomplishing this fiat. Gödel showed that any formal system strong enough to answer all questions in the system must contain an inconsistency, and that any system strong enough to do elementary number theory must have undecidable questions.

Around 1908 in the midst of the logicism formalism controversy, a mathematician by the name of Brouwer injected an idea which has become known as *intuitionism.* Brouwer maintains that one cannot separate inquiries into the foundations of mathematics from discussions of the conditions which produce mathematical activities. Thus, form alone tells us only of the language in which mathematics is expressed. In viewing logic, he maintains that mathematics is independent of logic. We must consider logic as dependent on mathematics. Thus, mathematics and its construction are independent of any logic or philosophic system. Consequently, mathematical objects and processes are judged on the basis of what is intuitively clear. A very important result of this is that the unrestricted use of the law of the excluded middle that played such an important role in the logic of Aristotle (recall from Chapter 4 that this is stated as: either a proposition or its denial is true) was curtailed.

Summary

It should be clear from many things we have said in previous sections that among the most important tools to be evolved in mathematics in recent years is that of a function and a relation. Thus, as an exercise in mathematical research, look in elementary-school mathematics textbooks (a college library should have an abundant supply) for all the ways in which the word "function" or "relation" is used. Repeat the same process for junior-high and high-school material. Finally, write a short report on the use of the concept of "function" or "relation" that would be useful in your particular subject-matter area. If you need help, ask and discuss these questions with other people. Mathematics can be a social experience.

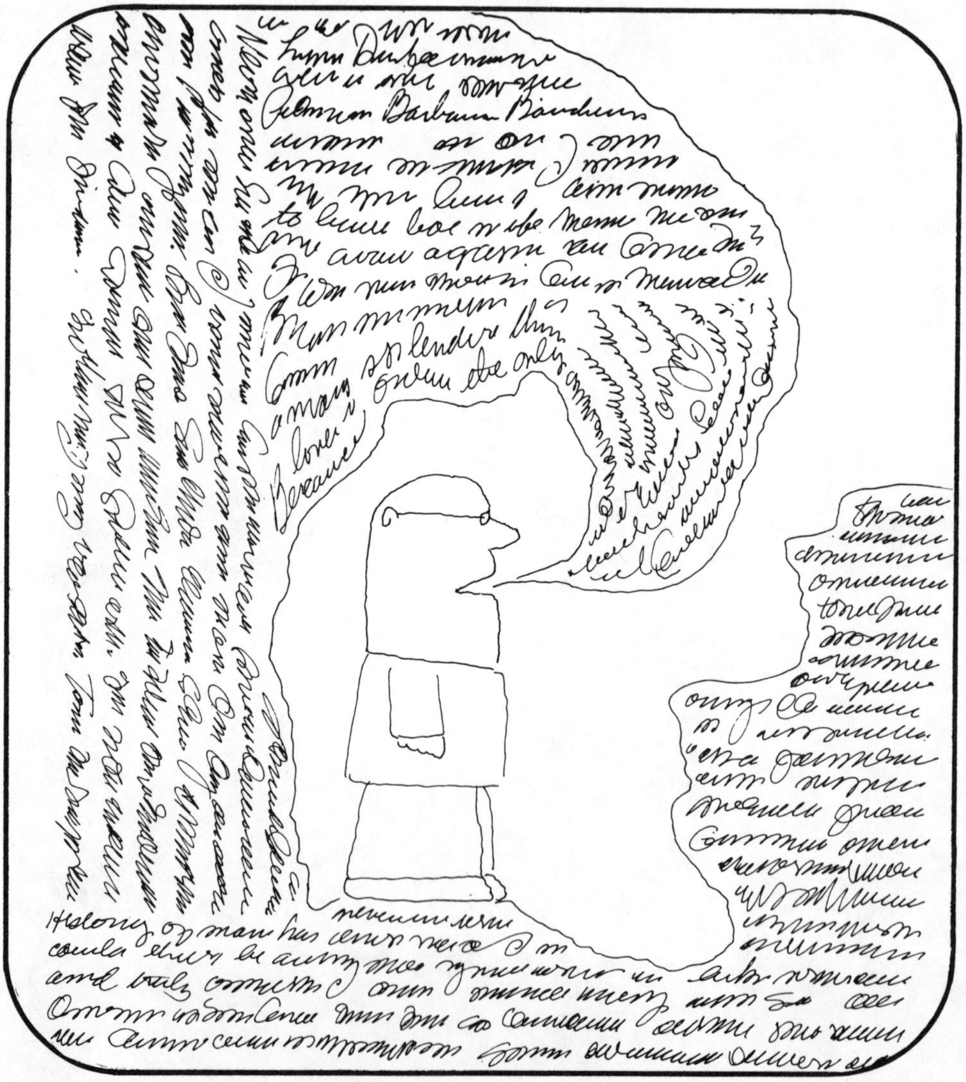

splendor
only
Because
History of man has
could
and truly

Chapter Six

Language and Automated Processes

"Twas brillig, and the slithy toves
Did gyre and gimble in the wabe,
All mimsy were the borogoves,
And the mome raths outgrabe."

Through the Looking Glass

Lewis Carroll

The previous parts of our discussion have been devoted mainly to considerations of thought systems, their associated logics, and experiences as they related to changes in perspective in the nature of the mathematical sciences. Because of the basic assumptions we made in Chapter 1, we now turn our attention to the concept of a language and its associated grammar.

What is a Language?

In one sense, it is quite easy to answer the question "What is a language?" A language is something with which to communicate (review Chapter 1). We wish to include in this primitive idea, not only communication among men, but also communication between machine and man and machine and machine. We wish to investigate those common properties of a communication vehicle that can in some natural way be termed as linguistic. Moreover, we wish to restrict the discussion to written forms.

For our purposes a language must consist of units we shall call sentences. Sentences can be further subdivided into subunits called phrases. Phrases in turn subdivide into smaller units called words which are assemblages of things from some alphabet. To be more accurate we should observe that not all written languages are structured in exactly this way. However, there are only two ways of denoting items in a language: phonetic characters and pictographs. Even these modes hardly ever occur in pure form.

Natural and Artificial Languages

One of the current research interests in mathematics and elsewhere is that of language. This is particularly true because mathematics has on numerous occasions been called a language and because mathematics itself must be communicated by a language. Therefore, some knowledge of language seems imperative.

For convenience, we shall place all languages into two categories: natural and artificial. Languages which have evolved as communication instruments of common discourse are *natural.* Examples of natural languages are English, German, French, and Chinese. *Artificial* languages, on the other hand, are those which have been purposefully invented to accomplish certain specific communication goals. Examples of these are mathematics and logic in certain of their forms. However, one of the best examples is that of computer languages. Discuss the idea that a natural language can be viewed as a set of artificial languages.

Before we proceed much further, we should mention that there seems to be two basic approaches to the study of a language. One might approach the subject from the standpoint of the humanities as a search for meaning, with form being a secondary concern. Our emphasis, on the other hand, shall be more concerned with form. We hope in the process that meaning will not be too distorted.

Grammar

One of the ways of specifying the structure, or form, of a language is to describe its grammar. Hence we need to know what a grammar is. Let us temporarily agree that a language consists of words, phrases, and sentences. There are now two paths we may take at this point. We may think of having a basic set of vocabulary words and using them according to certain prescribed rules to build phrases. Phrases, in turn, may be assembled according to certain rules to form sentences. The rules that accomplish this function might be termed a *generating grammar*.

On the other hand, we might begin with a sentence, try to identify phrases, and then identify words in each of the phrases. Each of these processes would be accomplished by a set of rules. This process could be accomplished by using a *parsing grammar*. In either case we see some common situations. To help us identify them, let us consider some examples.

Our first example will consist of a part of the English language. Suppose we have three types of words available: nouns, verbs, and articles. We have three types of phrases: subject phrases, predicate phrases, and direct-object phrases. The words *car* and *wall* are nouns. The words *a*, *an*, *the* are articles. We have but one verb, namely, *hit*. We shall stipulate the following phrase structure:

1. A **sentence** is a **subject phrase** followed by a **predicate phrase**.
2. A **subject phrase** is an **article** followed by a **noun**.
3. A **predicate phrase** is a **verb** followed by a **direct-object phrase**.
4. A **direct-object phrase** is an **article** followed by a **noun**.

Let us denote the above items as a set of "production rules."

1. **(σ) → (S) (Pr)**.
2. **(S) → (A) (N)**.
3. **(Pr) → (V) (DO)**.
4. **(DO) → (A) (N)**.

Since all of the symbols in parentheses serve as place holders or labels for the words we shall finally insert to make a meaningful sentence, they are sometimes called *syntactic variables*, or nonterminal vocabulary words. On the other hand, the words *car*, *hit*, *wall*, etc. are called terminal vocabulary. Let us now produce a real sentence from this. We shall generate this sentence by a sequence of applications of the above production rules.

(σ) → (S) (Pr)
→ (A) (N) (Pr)

→ **(A) (N) (V) (DO)**

→ **(A) (N) (V) (A) (N)**

→ *a* **(N) (V) (A) (N)**

→ *a car* **(V) (A) (N)**

→ *a car hit* **(A) (N)**

→ *a car hit the* **(N)**

→ *a car hit the wall*

Thus our "grammar" has produced a meaningful sentence. However, it is easy to see that a sentence such as

A wall hit the car

could have also been generated, using these rules. "Content" again is a stumbling block. According to our grammatical rules, this sentence is correct. The form, or *syntax*, is acceptable; however, the sentence is semantically undesirable. (Note that, intuitively, syntax pertains to form while semantics has to do with content or meaning.)

Although the same rules cannot be used to parse a sentence, a slight modification will permit us to parse a certain class of sentences (unfortunately with somewhat questionable results). Change the rules to read as follows:

1′. **(S) (Pr)** → **(σ)**
2′. **(A) (N)** → **(S)**
3′. **(V) (DO)** → **(Pr)**
4′. **(A) (N)** → **(DO)**
5′. *wall* → **(N)**
6′. *car* → **(N)**
7′. *a* → **(A)**
8′. *an* → **(A)**
9′. *the* → **(A)**
10′. *hit* → **(V)**

Thus we might try to parse the following sentence:

The car hit a wall → **(A)** *car hit a wall*
→ **(A) (N)** *hit a wall*
→ **(A) (N) (V)** *a wall*
→ **(A) (N) (V) (A)** *wall*
→ **(A) (N) (V) (A) (N)***
→ **(S) (V) (DO)**
→ **(S) (Pr)**
→ **(σ)**

However, beginning at line (*) something other than the rules is in action. In the very next line we proclaim that the first **(A)**(**N**) combination is (**S**). How do we know that? The answer is: "We don't!" The combination could just as well have been (**DO**). Of course, any moderately intelligent young student of the English language would be quick to observe that the latter would be *semantically* unacceptable. In other words, because our student is able to "understand" the meaning of the sentence, he knows never to choose (**DO**) as an alternative. The problem is: How does one devise unambiguous rules governing decisions of the above type that are completely independent of meaning? As we shall see, automated processes are capable only of syntactic and logic decisions.

Let us look at another example, this time one that is more capable of formalization. Remember items in parentheses are syntactic variables and in many cases a syntactic variable can be replaced by many items in a list. We shall use a vertical bar (|) to separate these items. Rather than using (→), let us now use (: : =) whenever a definition is intended. The arrow notation will still be used for one-way productions.

This is how to construct a numeral (integer or decimal).

(symbol)	: : = **0** | **1** | **2** | **3** | **4** | **5** | **6** | **7** | **8** | **9** | **.**
(digit)	: : = **0** | **1** | **2** | **3** | **4** | **5** | **6** | **7** | **8** | **9**
(integer)	: : = (digit) | (digit) (integer)
(decimal numeral)	: : = (integer) | (integer) **.** | (integer) **.** (integer)

We need to explain the last two definitions. The third one means that an integer is either a single digit or a single digit followed by a previously defined integer. This type of definition is called *recursive*. Thus objects such as 0, 5, 10, and 0001 can be considered integers. The fourth definition is like the third, for this says that a decimal numeral is either an integer or an integer followed by a decimal point or an integer followed by a decimal point followed by an integer. Thus 0.0001, 5, and 612.518 are all legitimate decimal numerals, but .0001 is not legitimate—according to the rules. (Why?) What difference if any would it make in the rule for (integer) if we write (integer) (digit) instead of (digit) (integer)?

Exercises

1. (alphabet) : : = *C* | *A* | *K* | *E* | *N* | *p* | *q* | *r* | *s*
   (bicon) : : = *C* | *A* | *K* | *E*
   (variable) : : = *p* | *q* | *r* | *s*
   (*wff*) : : = (variable) | *N*(*wff*) | (bicon) (*wff*) (*wff*)
   Decide within the rules whether or not the following strings of alphabetic characters are wff's
   (a) *p*
   (b) *pq*
   (c) *Np*
   (d) *Cpq*
   (e) *CNpq*
   (f) *CpNq*
   (g) *NNp*
   (h) *NNNNNNp*
   (i) *NKNpKqKNrs*
   (j) *CpKEpqANqNCpEqr*
   (k) *CAKEpNqNrNCAKEpqNqNpr*
2. (name) : : = Megan | Jill | Patrick | Mommy | Daddy
   (tattle verb) : : = Hit | scratched | bit | spanked

(parent) : : = Mommy | Daddy
(tattle tale) : : = (name) (tattle verb)(name) | (name) (tattle verb) me | (parent) (tattle tale)

a. Show that the following are tattle tales.
   (i) Jill hit Patrick
   (ii) Patrick bit Megan
   (iii) Mommy, Daddy spanked me
   (iv) Mommy, Mommy, Mommy, Mommy, Megan scratched me
   (v) Patrick hit Patrick
b. Make up some of your own, using the rules.

Let us now summarize the impressions motivated by the preceding material. It should be clear that a language is completely known once we know all the sentences of that language or at least know how to get them. This, of course, also means that for a given sentence we should be able to decide whether or not the sentence belongs to the language. This is what a grammar is supposed to do for us. Since it is clearly a set of rules for producing sentences, a language's grammar should ideally produce all possible sentences of the language. However, we would also like a grammar to produce only sentences in the language. The examples above seem to indicate that this might be somewhat difficult if not mathematically impossible. In fact, if you look closely, you might even see a relationship between this situation and the Gödel incompleteness theorem mentioned earlier.

Exercises

1. Write a short paper on how mathematics affects or could affect your subject-matter interests. (If you are a mathematics major explain what it might mean to think or converse mathematically about mathematics.)

2. Write a paper comparing syntax to semantics. In what ways do they depend on each other? In what way are they independent?
3. Devise a grammar to unambiguously parse "She is a pretty little girl."

Algorithm

The concept of an algorithm is one that is familiar to everyone, though the name may not be. At the very least an algorithm is a list of instructions to carry out some process in a step-by-step fashion. Recipes in cook books are excellent examples of algorithms. In grade school arithmetic we learn the basic operations with numbers by memorizing algorithms. Thus, the product of 325 times 15 can be accomplished by multiplying 325 by 5 and then multiplying 325 by 10 and adding the two results as given below:

$$\begin{array}{r} 325 \\ \times\, 5 \\ \hline 1625 \end{array} \qquad \begin{array}{r} 325 \\ \times\, 10 \\ \hline 3250 \end{array} \qquad \begin{array}{r} 1625 \\ +3250 \\ \hline 4875 \end{array}$$

Or, it can be arranged like this:

$$\begin{array}{r} 325 \\ 15 \\ \hline 1625 \\ 325 \\ \hline 4875. \end{array}$$

Alternately, one can invoke the definition of multiplication as repeated additions:

325
325
325
325

$$\begin{array}{r} 325 \\ 325 \\ 325 \\ 325 \\ 325 \\ 325 \\ 325 \\ 325 \\ 325 \\ 325 \\ \underline{325} \\ 4875 \end{array}$$

The ancient Egyptians had the following algorithm:

1. Halve one factor until 1 is obtained.
2. Double the other factor a like number of times.
3. Add the results from step 2 that correspond to odd answers in step 1.

Thus we might have:

$$\begin{array}{rr} 15 & 325 \\ 7 & 650 \\ 3 & 1300 \\ 1 & \underline{2600} \\ & 4875 \end{array}$$

or

$$\begin{array}{rrr} 325 & 15 & 15 \\ 162 & 30 & \\ 81 & 60 & 60 \\ 40 & 120 & \\ 20 & 240 & \\ 10 & 480 & \\ 5 & 960 & 960 \\ 2 & 1920 & \\ 1 & 3840 & \underline{3840} \\ & & 4875 \end{array}$$

The moral of these examples is that there may be more than one algorithm to accomplish the same task. Many times the device to execute a process must be taken into account in the construction of an algorithm. If we are telling another person how to do something many times we need not be quite as precise as we must be when telling a machine. With a person we can usually ask and answer questions. However, we shall see that with a machine this is not possible. But in either case an algorithm must be expressed in terms of certain fundamental operations, or elementary procedures, which it is assumed are meaningful to the devices used in execution regardless of whether they are people or machines.

Let us consider another example. This time let us borrow from high-school mathematics. The problem is to find all values of x such that

$$ax + b = 0,$$

where a and b are any real numbers. High-school algebra "solves" the problem in two easy steps:

1. Subtract b from both sides;
2. Divide both sides of the equation by a.

Now, as long as $a \neq 0$ this yields a nice usable solution $x = -(b/a)$. However, the algorithm said nothing about $a = 0$! Moreover, from where are a and b obtained? Let us try another approach.

1. Read value for a, read value for b.
2. If $a \neq 0$ then $x = -(b/a)$ type x else go to 3.
3. If $b = 0$, then type "any value of x works" else type "no x works."

   By way of summary we note the following have characteristics of an algorithm:

1. An algorithm is a set of instructions of finite length;
2. A device of some kind responds to the instructions and executes them;
3. A device controls the process allowing for execution, storage, and retrieval;

4. The procedure is step-by-step, that is, discrete;
5. At each step there is one and only one way of proceeding to the next step. There is no place for question or interpretation;
6. The process must terminate after a finite number of steps preferably by some instruction in the algorithm.

Exercises

Devise algorithms for the following processes.

1. How to take an examination.
2. How to tie a shoelace in a bow knot.
3. How to write a term paper.
4. How to look up a word in a dictionary.
5. How to convert base 10 numerals to any other base less than 10.
6. How to find the remainder when 196 is divided by 19.
7. How to convert CAT into TIN changing one letter at a time in three steps. At each step you must have an English word.

Flowcharts

In this section we shall present the concept of a flowchart. Flowcharts are pictorial, or graphical, representations of algorithms. They are many times used as an intermediary step between the statement of an algorithm in a natural language and its translation into a program language. A programming language is an artificial language acceptable by some device. It is one in which communication can take place between person and device or device and device (even person and person).

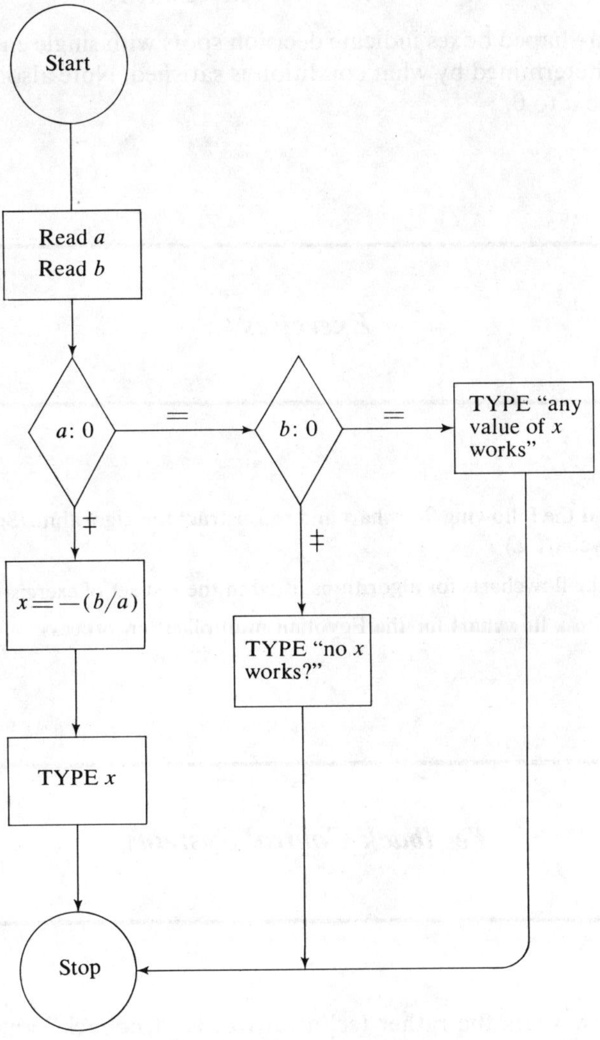

Figure 6-1. Flowchart 1

A flowchart consists of "boxes" which contain notation representing processing, input/output, operations and decisions, and directed line segments connecting the boxes with an indication of direction of flow. Thus we might draw a flowchart of the above algorithm for solving

$$ax + b = 0. \qquad \text{(See flowchart 1)}$$

The diamond-shaped boxes indicate decision spots with single entry and multiple exits. The exit is determined by what condition is satisfied. Note also that $a{:}0$ is to be read as "compare a to 0."

Exercises

1. Read the following flowchart and reconstruct the algorithm. (See flowchart 2.)
2. Make flowcharts for algorithms listed in the last set of exercises.
3. Make a flowchart for the Egyptian multiplication process.

Feedback Control Systems

In the past few years the rather technical words "feedback" and "cybernetics" have assumed a role even in everyday language usage. The word "cybernetics" was introduced by the mathematician Norbert Wiener.* The word cybernetics derives from a Greek word meaning helmsman. Thus, in a simplistic manner cybernetics means control. Wiener observed that certain machines and organisms behaved in certain similar fashions. He tried to lay down the rules for these similarities.

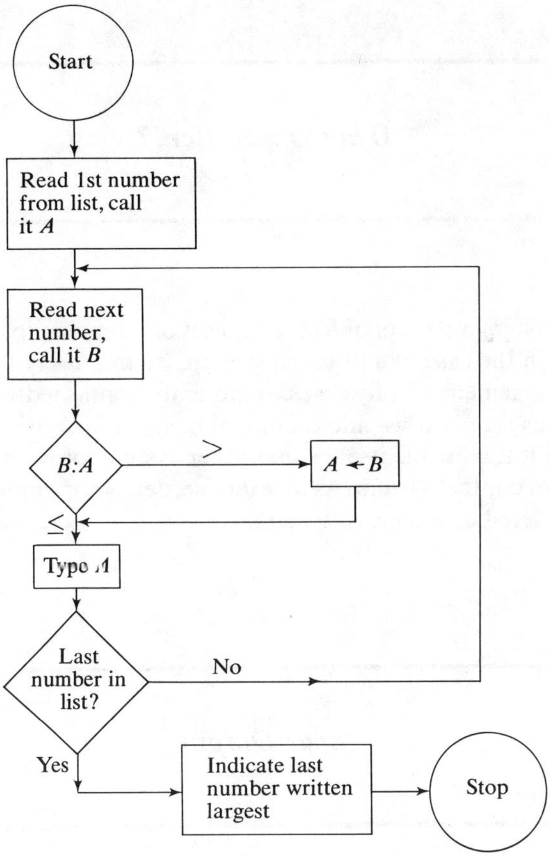

Figure 6-2. Flowchart 2

* See Norbert Wiener, *Cybernetics.* Cambridge, Mass: M.I.T. Press, 1948.

The idea basic to the process of control is feedback. It seems to be one of the most fundamental processes existing in nature. It is present in nearly all dynamic systems: within man and animal, in and among societies, within machines, and among machines as well as in man-machine systems. Feedback is best understood in reference to systems, so let us take a look at some important systems.

What is a System?

In general, a system is a set of objects and a set of relationships existing among the objects of the set. In the case of a physical system, we might say that we have a set of objects or an arrangement of physical components connected or related in such a way as to behave as a unit. If we add a control device to a system we have a *control system* if and only if it controls itself, some other system, or both. We shall later see that a computer is a control system. As an exercise, defend or oppose the position that logic can be considered as a control system.

Input/Output

Basic to the kinds of control systems with which we shall be interested are those which can be given information, data or stimuli of some sort. This information is supplied from external sources, and usually to obtain a desired response. Note the similarity between this and the communication network given in Chapter 1. This given information is called the *input* to the system.

The *output* of a control system is the response. Unfortunately, it may or may not be the desired one. As we saw earlier, one of the inhibitions to desired response, of course, could be *noise* in the communication channel.

Classification and Examples

There are several ways of classifying control systems. We shall present two ways here. The first type of classification is as follows:

1. Man-made systems;
2. Natural systems;
3. Hybrid systems.

Hybrid means a combination of man-made and natural systems.

Let us look at the following examples.

1. (a) An electric switch.
   (b) A thermostatically controlled heater which regulates the temperature of a room or other enclosure such as an oven.
2. (a) The system consisting mainly of the eyes, an arm, hand or finger, and brain of a man when pointing at something.
   (b) Perspiration system of an animal.
3. (a) A man driving a car.

As a trivial exercise, classify each of the examples as man-made, natural, or hybrid.

A second way of classifying control systems is by classifying them as open and closed systems. These are sometimes called open-loop and closed-loop systems. An open-loop control system is one in which control is independent of output. An example of this type is an automatic toaster controlled by a preset timer. On the other hand, a closed-loop control system is one in which control action is dependent on output. Examples 1(b) through 3 are examples of closed-loop systems.

Feedback then emerges as a property of closed-loop systems and helps to distinguish them from open-loop systems. Examine the Figure 6-3.

Figure 6-3

Notice that some input stimulus is received by a receptor device, such as an eye. A message is then sent to a control unit (brain). This in turn sends a message to an appropriate effector such as an arm, hand and finger combination. The effector action, in addition to providing an output, is relayed back as an input to the receptor.

Hence, feedback is that property of a closed-loop system which permits the response of the system to be compared with the input so that control of the system is a function of both input and output.

Automated Action

We are not trying to make computer scientists of you. We will not provide you with knowledge of the intricacies of programming a computer. Rather, our purpose is to give you a glance at a conceptual model of a computer and some of the problems involved in communicating with a computer. This material draws together many of the preceding sections.

To build our conceptual model of a computer we shall begin with the idea of an automaton. An *automaton* is a device which can accept inputs, process them in some way, and produce outputs. In general an automaton will have various internal states in which it may be. Along with the input these states determine the output. Moreover, an input into a given state of the device determines the state in which the device will be for the next input; that is, the next state of a machine is a function of the present input and state. Many devices, therefore, are automatons. Thus, the thermostat mentioned earlier as well as traffic-signal lights are automatons. A computer can be thought of as an automaton designed to perform certain mathematics-like operations on data consisting of symbols.

There are three kinds of computers of interest to most of us today. The *analog computer* represents numbers by essentially *continuous* physical things or quantities. A slide rule and thermometer are examples of analog computers. On the other hand, the type of computer we are most interested in is one in which numbers and symbols are represented by *discrete* things like digits, switches, or beads. These computers are called *digital computers*, or *data processors*. Hence, one's fingers and toes as well as such exotic devices as an abacus are examples of a digital computer. Today, it is not uncommon to find a digital and an analog device hooked together so that the features of each can be used. Computers of this type are called *hybrid*.

However, the device we shall be most interested in is a digital computer, which has the added feature of being a stored-program device. Figure 6-4 shows that a digital computer obviously has the property of being a feedback-control device.

Figure 6-4 is largely self-explanatory; however, let us use this to summarize. Just like all control devices, a computer has ways of receiving data usually through punched cards, typewriters, or in some instances voice commands. It has an output device which either prints messages, punches cards, or in some other way outputs information. It has a memory in which information can be stored and retrieved. Actually many computers have at least two memory sections. One retains information until needed, a kind of permanent memory. The other works with that information. Later we shall see how this might work in conjunction with the processing unit in which the work of the machine is carried on. This unit has the ability to follow instructions and perform a variety of operations, including storing in memory, fetching from memory, addition, subtraction, multiplication, division, and comparison. The processor in conjunction with the control unit can carry out lengthy sequences of prescribed plans. It can, in fact, make "decisions" based on its own work and therefore take alternate paths through its instructions.

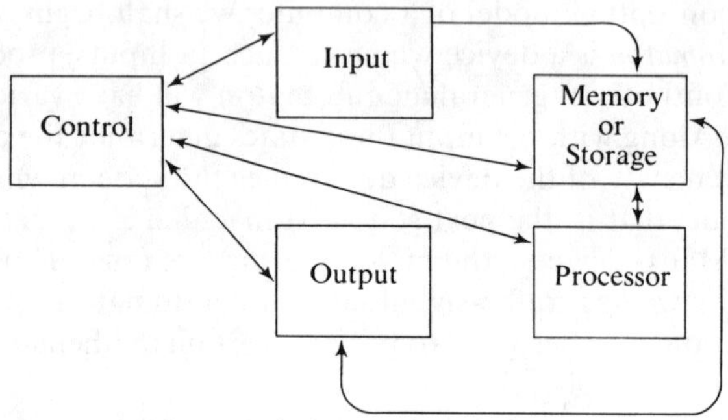

Figure 6-4

A Conceptual Model

In our last section we defined the type of computer in which we were going to be primarily interested. We also designated its parts according to function. In this section we will build a conceptual model of a part of a computer. In this way we shall be better able to understand some of the functioning of the computer itself. However, as an added benefit we hope that you will discover some interesting things about the way you think and process data. We mention in passing that in many ways an electronic computer operates in a fashion similar to that of the neural nets that compose your brain. This is true in the following sense. The brain is composed of individual elements called neurons; these are either firing or not firing. The working part of a computer is composed of tiny rings of metal that can be magnetized in one direction or another. (See Figure 6-5.) Thus, both situations can be represented by binary numbers denoting "on"–"off" conditions. Moreover, it is easily seen that both a brain and a computer share a basic common language composed entirely of strings of

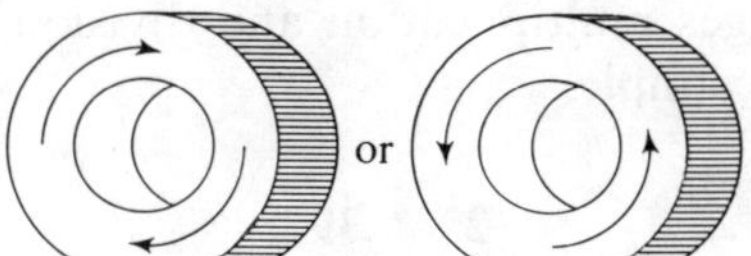

Figure 6-5

0's and 1's. Unfortunately (or fortunately depending on your own view) these languages are different and are not readily understood by any of us. In fact, it is doubtful that any individual knows his own brain language or could even learn it. It is enough that it operates. We feel the same way about machine language. We just know that it's there.

Just as we must know what languages other individuals speak in order to communicate with them, we must find suitable computer languages in order that man-computer discourse may take place. How much and what kind of language are needed depends mostly on one's aims. Many of you need only become proficient in how to flowchart because you will communicate through programmers. Others need to become proficient in one or more of the sophisticated languages in existence or even in the languages that the machine's basic circuits use.

It should not be too surprising when I tell you that a conceptual model of a computer can be developed around the same three things that we based our conceptual model of a human being's worlds on. Primarily, we shall focus upon the relationship between computer language and computer-thought systems.

Similar to human written language, computer languages begin with an alphabet such as

A, B, C, D, E, F, G, H, I, J, K, L, M, N, O, P, Q, R, S, T, U, V, W, X, Y, Z

0, 1, 2, 3, 4, 5, 6, 7, 8, 9, +, −, *, /, =, (,)

In addition we shall use a period ".", a comma ",", and a single quote " '."

The statements most basic to any computer language are the arithmetic statements such as

$$2 + 3,$$

$$2 - 3.$$

In most computer languages multiplication and division are symbolized by using * and /, respectively. For example,

$$2.3 * .392,$$

$$1.956/5.$$

Parentheses play their traditional role as separators:

$$(2.3) * (-60.8)/(2 + 6.7).$$

The symbol "=" plays a nontraditional role. To best understand its use, we introduce the idea of a *label.* A label does two things for us. First, a label names a number or a set of numbers. Second, a label sets up a place in the memory of the machine. Now the memory of the machine can be pictured as pigeon holes whose designation can be changed with need. These holes are thought of as containing one number when in use. One may view this number at will without disturbing the contents. One may replace the number with a new one without relabeling.

Now let us see how to construct a label. For our purposes a label is a string of alphabetic or numerical characters from one to five characters in length. The first character must be alphabetic. Thus the following are admissible labels:

1. A
2. AB
3. A1
4. Z795
5. LOGIC

As an exercise tell which of the following are admissible as labels:

1. 2B4
2. ALGAE
3. TOM
4. ANY
5. FIVE
6. TWELVE

Actually all of the above except numbers 1 and 6 are acceptable as labels. Hopefully, you know a label when you see one. Let us determine the meaning of such statements as

1. A = 3.1, B = 5, C = 5.
2. B = B + 1.
3. C = 2 * C + A/B.

Let us for the moment assume that A, B, C are labels. Then line one says "find three places in memory, label them A, B, C respectively, put 3.1 into that place called A; put 5 into that place called B; and put 5 into the place called C."

On the other hand, line two says "Take the number presently in the place called B, add a 1 to it and put the result back into B." Finally, statement number three instructs the machine to "Take the number stored in C, multiply it by 2. Take the number in A and divide it by whatever is in B. Add this quotient to that number which you obtained by multiplying 2 times C." And, store this result back in C. Thus "=" has a different meaning than is usually given it.

Some Special Strings

Many computer languages with a few modifications share the above type statements. We wish now to begin to examine one in particular. It is called AMTRAN. There are at least two versions of this language. One of these was developed at Brown Engineering Company, Huntsville, Alabama. The other version was developed at the University of Georgia Computing Center, Athens, Georgia. We will concentrate on the former version. Both were designed for use on an IBM 1130 computer.

In all computer languages certain combinations of symbols always have fixed meaning. The first of these in AMTRAN is the string

TYPE

The following examples show the usage of the word TYPE:

```
1. TYPE 2.
2. TYPE 2 + 2.
3. A = 1, B = 6, TYPE A + B.
4. A = 1, B = 6, TYPE (A, B).
```

For line one the computer will type 2. In line two it will type 4. In line three it will type 7. In line four it will type out two numbers: 1 and 6.

Now let us suppose we have a sequence of instruction that we give a computer, in other words, a *program*.

```
1. AT = 3, BA = 5.
2. CAT = AT + BA.
3. DOG = AT * CAT.
4. AT = AT + 1.
5. TYPE CAT, TYPE DOG.
```

The computer will type out the numbers

8 and 24.

Make sure you know why before proceeding.

Among the many powerful features of a stored-program digital device is its ability to repeat already given instructions as many times as desired. One way of accomplishing this feat is by use of an instruction such as

GO TO *n*.

The *n* is the number of the first instruction that one desires to be repeated. In this particular dialect of AMTRAN we can only refer to an already given instruction. Thus, suppose we have

1. AT = 3, BA = 5.
2. CAT = AT + BA.
3. DOG = AT * CAT.
4. AT = AT + 1.
5. TYPE (CAT, DOG).
6. GO TO 2.

Upon receiving the unconditional branch instruction

GO TO 2.

the processor reexecutes statement 2 and *all* statements following 2. We thus obtain the following values for

CAT	DOG
8	24
9	36
10	50
11	66
12	86
etc.	etc.

Because this type of branching is unconditional and because we have not written an instruction telling the computer when to stop, it will continue to *loop* until some manual termination method is used.

Repeat n

If we know exactly how many times we wish to repeat a given process, the AMTRAN language provides us with an instruction that we may give the machine. Thus the above program could be rewritten as follows if we wish to have only five repetitions.

1. AT = 3, BA = 5.
2. REPEAT 5, CAT = AT + BA, DOG = AT * CAT, AT = AT + 1, TYPE (CAT, DOG).

This program will output five values for CAT and five values for DOG. As an exercise tell how you might change this program to print only the fifth value of CAT and DOG.

Conditional Branching

As in the case of the algorithmic language and flowchart language, computer languages such as AMTRAN provide a statement enabling decision to be made so that certain steps in the process may be executed and others skipped, depending on circumstances at the moment of testing. In AMTRAN, this takes one of two forms. One of these is

IF ____________ THEN ____________.

The first blank is filled with a statement which is either true or false. Since most of our work involves comparing two quantities, these are the kinds of statements which are admissible. What are the possibilities when comparing two quantities, A and B? One of the following can occur:

1. A is less than B;
2. A is less than or equal to B;
3. A is greater than B;
4. A is greater than or equal to B;
5. A is equal to B;
6. A is not equal to B.

In AMTRAN, these so-called *Boolean* statements are translated respectively as:

1. A LT B
2. A LE B
3. A GT B
4. A GE B
5. A EQ B
6. A NE B.

These statements are called Boolean because inside the computer they are assigned the value 1 if true, and 0 if false. This was Boole's contribution to the arithmetizing of logic.

The second slot can be filled with any legitimate AMTRAN statement. The statement following THEN is executed when the Boolean statement is true. In the event the Boolean statement is false, the computer waits for further instructions.

Examples: We shall always begin a sentence of instructions by ENTER PROGRAM followed by a numbered list of instructions called the *program.*

1. A = 3.
2. IF A LT 10 THEN TYPE A + 3.

The computer here will return 6 as output. On the other hand suppose we have:

ENTER PROGRAM

1. A = 3.
2. IF A EQ 10 THEN TYPE A + 3.

The computer in this case will wait for a new instruction. Hence, suppose we have:

ENTER PROGRAM

1. A = 3.
2. IF A EQ 10 THEN TYPE A + 3.
3. B = TYPE a + 8, TYPE B.

Then from the computer we would get back

11

as output since TYPE A + 3 is ignored.

Sometimes it is more convenient to combine statements 2. and 3. in the last program. This can be done using an alternate form of the IF statement.

IF____________THEN____________ELSE____________.

Thus, the above becomes:

```
ENTER PROGRAM
1. A = 3.
2. IF A EQ 10 THEN TYPE A + 3 ELSE B = A + 8, TYPE B.
```

Let's try to read the following program:

```
ENTER PROGRAM
1. A = 3, B = 5, N = 1.
2. C = A + B, D = A * C.
3. IF N EQ 10 THEN TYPE (C, D) ELSE A = A + 1, N = N + 1,
   GO TO 2.
```

The output of this sequence of instructions is

17 204

Explain why.

Now let us write a program that asks for a fixed number of numbers, totals them, and then divides by the number of numbers provided.

```
ENTER PROGRAM
1. INPUT N, I = 1, S = 0.
```

2. INPUT X.
3. S = S + X.
4. IF I EQ N THEN A = S/N, TYPE S, TYPE A, GO TO 1
   ELSE I = I + 1, GO TO 2.

Let us trace the action carefully. Upon typing

1. INPUT N, I = 1, S = 0.

the machine will print a statement asking you for a value for N. It writes

ENTER N.

whereupon we might enter 2. We now type

2. INPUT X.

The machine types

ENTER X.

which we do, say, 10. Now we type

3. S = S + X.

Thus, after this introduction, S is 10.

4. IF I EQ N THEN A = S/N, TYPE S, TYPE A, GO TO 1
ELSE I = I + 1, GO TO 2.

Since I is currently 1, then I EQ N is false. Thus, the THEN statement is skipped, and the ELSE part is done, so 1 is added to I, and we return to statement 2, which asks us to input another value for X. Let's type in an 8. This is added to the 10 in S to give 18. I is now 2 which is the same as N, so I EQ N is true, and we have what follows the THEN done:

A = S/N	which is 18/2 = 9
TYPE S	which is 18
TYPE A	which is 9
GO TO 1	

The last part of the statement permits us to choose another N

Exercises

1. Explain what the following program does overall. Then explain each step.

ENTER PROGRAM.

1. INPUT EMNO, INPUT HOURS, INPUT RATE.
2. IF EMNO LT 50 THEN WAGE = HOURS * RATE, TYPE (EMNO, HOURS, RATE, WAGE), GO TO 1 ELSE WAGE=HOURS * RATE, TYPE (EMNO, HOURS, RATE, WAGE), TYPEOUT 'END OF PAYROLL'.

2. Explain the following program.

ENTER PROGRAM.

1. N = 1.
2. A = N * N, B = A * N, C = SQRT N.
3. IF N EQ 10 THEN TYPE (N, A, B, C) ELSE TYPE (N, A, B, C), N – N + 1, GO TO 2.

Note: SQRT N means take the square root of N.

3. Write a program to find the average of ten numbers, and find the standard deviation. Take two sets of ten numbers, and write a program to compute the coefficient of correlation. These are statistical terms, so if you do not understand their meaning, look them up in an elementary statistics book.
4. Write a program to compute the quotient and remainder upon division of one number by another. Use the fact that division is repeated subtraction.
5. Write a program to change a numeral base 10 to a base *n* where *n* is some number 2 or larger, but less than 10.
6. Try to make up problems of your own and try to write programs to compute the solutions.
7. Analyze the use of the comma as it is presented above.

Postlude

In preceding chapters I have tried to share a number of things with you. First, I have tried to demonstrate that there are things other than numbers with which mathematicians concern themselves when they function, or claim they are functioning, as mathematicians. It should be apparent that mathematicians doing mathematics are concerned with the analysis and synthesis of structure. Thus, few if any subject-matter areas are free from the invasion of mathematics and mathematicians.

Of course, I don't think of this intrusion as dangerous. Rather, I think it can be healthy. First, the mathematician is provided with new objects for study. Second, areas outside of the mainstream of mathematical activity might be provided with tools and attitudes to solve problems with which they have been faced and which might have seemed completely insoluble.

As observed earlier, I have tried to avoid detailed study of arithmetic, algebra, and geometry as such, and other traditional first year college mathematical subject matters. I have tried to concentrate on what I called a mathematical attitude—everything that exhibits a structure or might exhibit a structure has been treated as an object of mathematical study. Notice that even in talking about mathematics I tried to be mathematical. In most cases undefined terms were explicitly given as such. I tried to be as careful as possible about stating axioms as well as definitions. Of course, in informal exposition of this type, formal rigor has been sacrificed to stylistic rigor.

Thus, no mathematician could be totally happy with the quality of mathematics presented. On the other hand, I am not writing for professional mathematicians, except incidentally.

A word or two on content perhaps is in order. Any author is prejudiced, and any choice of topics reflects to some extent his way of thinking about things. Several reports concerning first-year college mathematics courses have emphasized the necessity of opening doors to future mathematical studies. It seemed to me that traditionally we have assumed that the average student, by mysterious osmotic processes, will know what the doors are. My experience tells me this isn't the case. Hence, some introductory course should supply this need.

In most areas of study there is a vast literature for the lay person seeking edification. This is also the case in mathematics. Why not share this knowledge in a beginning course? Who knows, some might even express mild interest or even excitement. One of the most gratifying experiences I have had is when a student came into the course with the attitude "I dare you to teach me anything," and he discovered he could not only prove a theorem, but that he had the creativity to discover theorems of his own. Imagine if you will the shock of a female elementary education major who has hated mathematics because it was unladylike who suddenly discovers that a computer, and programming it, can be a fun thing with which to play.

I feel I must make some comment about some of the terminology, particularly that introduced in the first chapter. In trying to clarify these ideas for myself, I kept coming back to the mathematical constructs of category and topological manifold. While I quite frankly prefer the concept of category here (indeed, it seems far more natural), there was something useful to be served by introducing the word manifold into a context where a general reader might not expect to see it. Besides, using the word "category" for some people might call up all sorts of Kantian visions which may or may not be justified particularly at this early stage. I took heart, however, in the knowledge that the manifolds I was interested in could be thought of as a category!

The subjects I chose to include are debatable of course. I certainly lay no claim to superiority of choice over any other. I tried to think of the topics I would cover in a course of study for an undergraduate major. From these topics I tried to abstract and distill the principal ideas into a palatable form for the general public, developing what might be called "conversational mathematics."

Finally, let me state one of the ideas which has emerged in the course of these pages. It should be clear that a system of the formal logistic type is a quadruple of sets, $S = (\mathscr{A}, \mathscr{X}, \mathscr{F}, \mathscr{D})$, such that:

1. $\mathscr{X}$ is a finite or denumerable alphabet;
2. $\mathscr{F}$ is a language over $\mathscr{X}$. In particular, $\mathscr{F}$ is a formal artificial language and $\mathscr{F}$ is called the set of formulae for $\mathscr{S}$.
3. $\mathscr{A}$ is subset of $\mathscr{F}$ called the axioms of $\mathscr{S}$.
4. $\mathscr{D}$ is a set of deduction rules. That is a logic as we have previously defined it.

An Added Bonus—
Sets and the Single Student

In a very few pages we shall try to give you a brief outline of what is sometimes called naive set theory. It is naive in the sense that it is very intuitive (nonaxiomatic). More important, things will be written without regard to the difficulties that can occur where no axiomatic structure is imposed. We note in passing that if we were to take an axiomatic approach to set theory, we might take the word set as being undefined. The axioms then would be chosen in such a manner as to characterize completely the word set and other associated words, such as "element of a set." The mathematically mature will realize that there might possibly exist several axiomatic set theories as there exist several axiomatic geometries.

Conceptually, the word "set" symbolizes the idea of collections of individual things. The latter are called the *elements* of the set. For every given set, it must be possible in principle to decide whether a particular individual is or is not a *member* of the given set. Consequently, there are only two possibilities for any item, either it belongs to a given set or it doesn't. In the former case, the item is said to be a member, or element, of the set.

In view of the last paragraph, a set is specified by determining the composition of its membership in some way. Hence, a set is "known" when its members are either completely listed, some rule for listing is given, or certain propositions are given about the elements. In the latter event, if an item satisfies the condition of the proposition then the item belongs to the set specified.

The following is a complete listing of elements of a set which we designate by

$$A = \{1, 2, \text{pig}, \text{cow}, a, b\}.$$

As long as a set has a "finite" number of elements, we have no trouble, in theory, listing all the elements, though, in practice, this might be somewhat difficult. Thus, to write down all the integers from 1 to 1,000,000, though possible, would be laborious. Mathematicians, being somewhat lazy, denote this listing by the following

$$B = \{1, 2, \ldots, 1{,}000{,}000\}.$$

The pattern, he feels, is established by listing the first two elements and the last element while the three "dots" indicate that intermediate items are implicit, but not listed explicitly. Thus, the meaning of the three dots is "and so forth."

The above set also can be determined as follows. The set B is the set of all those integers n such that n is an integer between 1 and 1,000,000 inclusively. This is conveniently denoted by

$$B = \{n : n \text{ is an integer and } n \text{ is between 1 and 1,000,000 inclusively}\}.$$

Since the phrase "is an element of" occurs so frequently, this is denoted by $\in$. Thus, to say "x is an element of A," we would write "$x \in A$."

Let us look at some further examples:

1. $A = \{x : x = 0 \text{ and } x \leq 1\}$. Observe that $\frac{1}{2} \in A$, but $\frac{3}{2}$ is not an element of A. This we denote by $\frac{3}{2} \notin A$ or $N(\frac{3}{2} \in A)$ where N denotes "it is not the case that."
2. $Q - \{x : x \text{ is a rational number}\}$. A rational number is usually defined as any number which can be expressed as the quotient of two integers (cf. Chapter 2's discussion on $\sqrt{2}$). Which of the following are "true" propositions? Explain your answers.
   (a) $\frac{1}{2} \in Q$
   (b) $\sqrt{2} \in Q$
   (c) $1.5 \in Q$
   (d) $1.4999\ldots \in Q$
   (e) $a \in Q$
   (f) $N(2 \in Q)$

Set Relations

Two sets A and B may be related. We list the various possibilities:

1. A and B may have no common members;
2. A and B may have common members.

The two sets are said to be disjoint in the former case. The latter case provides several cases to be explored.

2(a). All of the members of A are members of B. Here we say A is a subset of B, and denote the situation by $A \subset B$. We have a similar situation if all the elements of B are elements of A. ($B \subset A$).

2(b). It may happen that only some of the elements of A are elements of B.

2(c). A and B could have exactly the same elements. In this event A and B are just different names for the same set, so we write $A = B$. We say that the two sets are equal. (Note that $A = B$ in this case if and only if $A \subset B$, and $B \subset A$ simultaneously.)

As an exercise classify the following pairs of sets according to the above relationships.

(a) $A = \{a, b, c\}, \quad B = \{1, 2, 3\}$

(b) $A = \{a, b, c\}, \quad B = \{a, b, c, 1, 2, 3\}$

(c) $A = \{x : x$ is an integer between 1 and 5 inclusively$\}$,
$B = \{1, 2, 3, 4, 5\}$

(d) $A = \{$pig, cow, John, Mary, hit$\}$, $B = \{$pig, John, Tim, big$\}$

Some Set Algebra

G. Boole and A. DeMorgan in the first half of the nineteenth century decided that sets could be combined in much the same way that numbers can. Many applications of set theory sprang from this idea. We shall investigate only two of these. First, however, let us try to see what Boole and DeMorgan had to say.

Let us make two assumptions at the outset: (1) There is a set of which all the sets in which we shall be interested at a given time are subsets. This set we call a *universal set*, or simply a universe; (2) We shall also assume there is a set which is a subset of every set in the above universe. This set, in particular, has no members. It shall be called an *empty set*. A universal set we shall denote by $\mathscr{E}$; an empty set we shall denote by $\emptyset$.

Let A and B be sets. We define the *union* of the two sets to be a set of all the elements of A plus those of B. We denote the union of A and B by $A \cup B$. A little reflection upon the matter will show that $x \in A \cup B$ if $x \in A$, or $x \in B$, or both. Thus, if $A = \{1, 2, 3\}$, $B = \{a, b, c\}$, then

$$A \cup B = \{1, 2, 3, a, b, c\}.$$

Similarly, if $A = \{1, 2, 3\}$ and $B = \{3, 4, 5\}$, then

$$A \cup B = \{1, 2, 3, 4, 5\}.$$

Notice also that for any set A,

$$\mathscr{E} \cup A = \mathscr{E},$$

and

$$\emptyset \cup A = A.$$

In general, if $A \subset B$, then

$$A \cup B = B.$$

Intersection

Let A and B be sets. We define their *intersection* to be the set denoted by $A \cap B$, consisting of all elements common to A and B. Hence, $x \in A \cap B$ if and only if $x \in A$ *and* $x \in B$. For example, if

$$A = \{1, 2, 3\}, \qquad B = \{a, b, c\},$$

then

$$A \cap B = \emptyset.$$

On the other hand, if

$$A = \{1, 2, 3\}, \qquad B = \{3, 4, 5\},$$

then

$$A \cap B = \{3\}.$$

Note that for any set A

$$\mathscr{E} \cap A = A, \qquad \emptyset \cap A = \emptyset.$$

In general, if $A \subset B$, then

$$A \cap B = A.$$

Difference and Complement

Let A and B be any pair of sets. The *complement* of A *relative* to B is the set of all those elements of B which are not A. We denote this by $B - A$. In the particular case when $B = \mathscr{E}$, we write A' instead of $\mathscr{E} - A$. By way of example, let $A = \{2, 3, 4\}$ and $B = \{1, 2, 5, 6\}$, then

$$B - A = \{1, 5, 6\}, \qquad A - B = \{3, 4\}.$$

Notice, therefore, that $B - A$ in general does not equal $A - B$. It is, however, easy to show that $A \cap B = B \cap A$, and $A \cup B = B \cup A$.

There is a sense in which $A - B$ can be thought of as a type of subtraction process. There is, however, another way to obtain set differences. We define the *difference* of two sets A and B to be the set $A \oplus B = (A - B) \cup (B - A)$. Thus, using A and B above [i.e., $A = \{2, 3, 4\}$ and $B = \{1, 2, 5, 6\}$], we see

$$A \oplus B = \{1, 3, 4, 5, 6\}$$
$$= (A \cup B) - (A \cap B).$$

Euler–Venn Diagrams

The above definitions are more readily seen and understood if we can draw pictures to model the definitions.

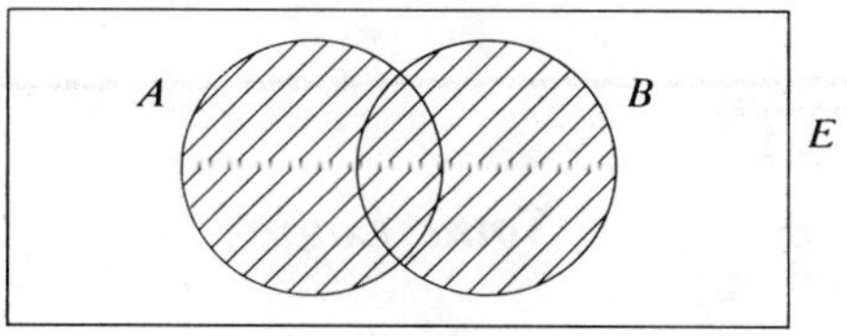

$A \cup B$

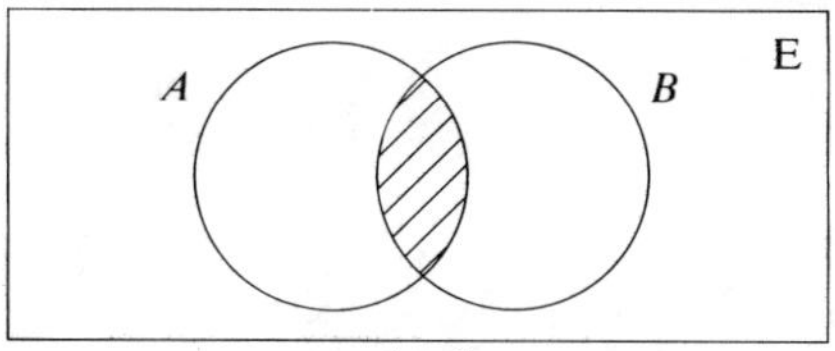

$A \cap B$

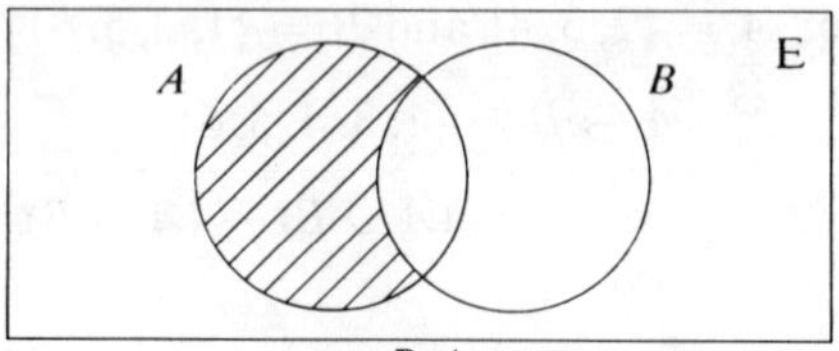

$B\text{-}A$

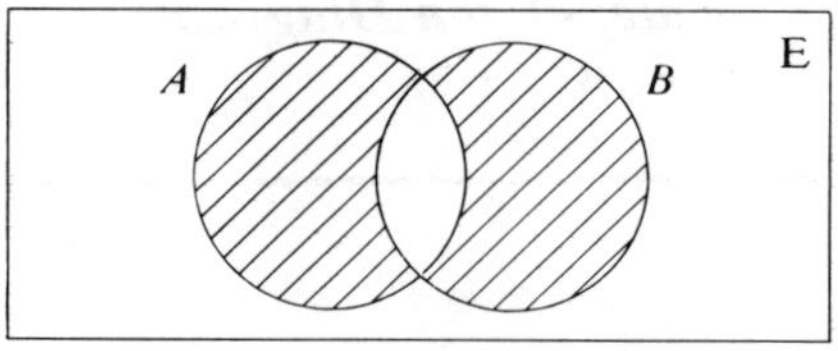

$A\phi B$

In each of the above cases the shaded area represents the set indicated.

Some Logic

In elementary logic there are usually four types of propositions studied:

1. categorical,
2. conditional,
3. alternative,
4. conjunctive.

The categorical proposition usually comes in four varieties. We illustrate these by:

1. All mathematicians are lazy people;
2. Some mathematicians are lazy people;
3. No mathematicians are lazy people;
4. Some mathematicians are not lazy people.

These sentences can be expressed by using the material on sets developed above. For example, the first sentence says, "the set of all mathematicians is a subset of the set of all lazy people," while the third sentence says, "the set of all mathematicians and the set of all lazy people have no common elements."

Now let

$$M = \{x : y \text{ is a mathematician}\},$$

$$L = \{y : y \text{ is a lazy person}\}.$$

We may translate each of the above sentences as follows:

1. $M \subset L$, or $M \cap L = M$, or $M \cup L = L$.
2. $M \cap L \neq \emptyset$.
3. $M \cap L = \emptyset$.
4. $M - L \neq \emptyset$.

These observations are especially useful when trying to put propositions together in the form of an argument. Classical logic studies situations such as "is it correct to deduce that all mathematicians are clods if we know that (1) all mathematicians are lazy people, and (2) all lazy people are clods?" Our ancient logicians said yes to this, and dreamed up elaborate reasons for explaining the whys and wherefores of similar *correct* or *incorrect* arguments. We usually refer to "correct" arguments as being valid and those which are "incorrect" as being invalid.

However, let's use our knowledge of sets to examine for validity or invalidity. Let

$$M = \{x : x \text{ is a mathematician}\},$$

$$L = \{x : x \text{ is a lazy person}\},$$

$$C = \{x : x \text{ is a clod}\}.$$

Then our sentences

1. All mathematicians are lazy people;
2. All lazy people are clods;
3. All mathematicians are clods;

translate as follows:

1. $M \subset L$,
2. $L \subset C$,
3. $M \subset C$.

Now, if subsets are to behave well with respect to inclusion, one would expect that from $M \subset L$ and $L \subset C$ we can obtain $M \subset C$. Notice that the correctness of the argument is independent of the truth or falsehood of any of the statements. This leads us to suspect that correctness of form depends on rules of operation and not on truth. Thus, validity means following the rules, and truth has a somewhat doubtful role.

What is truth's role? To determine this, let us examine another example.

1. All cows are pigs;
2. All pigs are four-legged animals.

After letting

C = cows,

P = pigs,

F = four-legged animals

we know that $C \subset P$ and $P \subset F$. We now obtain from these statements that $C \subset F$, which translates into "All cows are four-legged animals." This of course, using the usual meaning of the words, is true and valid in spite of the fact that (1) is false. Strange as it may seem, we can correctly *infer* something true from something false. Now observe:

1. All cows are pigs;
2. All pigs are sheep.

Following the above example it does not take much to see that from these items we may infer that

All cows are sheep.

This is slightly less unnerving for here we have falsehood being correctly inferred from falsehood. (See the discussion concerning the truth value table for "only if" in Chapter 4.) Let's examine a final example in this vein.

1. All pigs are animals.
2. All animals are made of hydrocarbons.

From these two "true" statements we are able to infer correctly that

All pigs are made of hydrocarbons.

This is certainly the case! Thus we see that from truth we can obtain truth by correct reasoning; however, by correct reasoning from falsehood we may obtain either truth or falsehood! Question: Do you think by using the set inclusion idea so far presented it would be possible to obtain falsehood from truth? How would you feel about a logic that would permit such a thing to happen? Carefully explain your answers. (Again refer to Chapter 4.)

Pathology Strikes

First, let us agree to dispense with the verbiage of the above examples since they show that our immediate concern is not content but form. Thus, our four types of categorical propositions become the following propositional forms:

1. All A is B;
2. Some A is B;
3. No A is B;
4. Some A is not B.

Now we are free to *define* the *undefined* A's and B's in any way we please. So suppose we know that

1. All A is B;
2. Some B is C.

What, if anything, can we deduce from these things about A's relationship with C? Unfortunately, one can say nothing with certainty, for either of the below situations is possible. Thus we have ambiguity, not certainty.

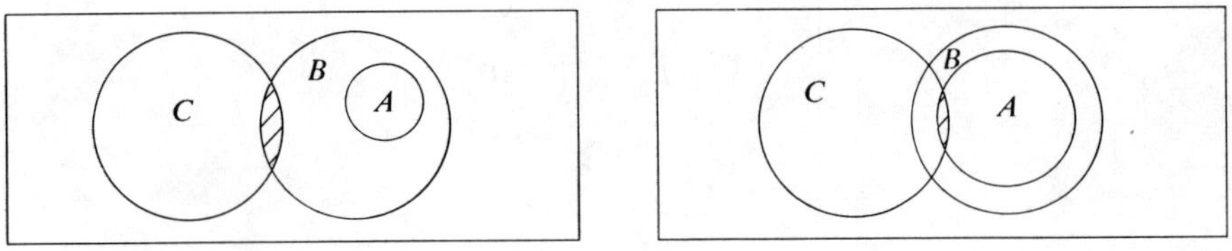

Most important, content would not *in general* help to clarify the situation. Consider:

1. All A is B;
2. No B is C.

What if anything can we conclude? Let's translate to sets:

1. $A \subset B$;
2. $B \cap C = \emptyset$.

Or diagrammatically, we have:

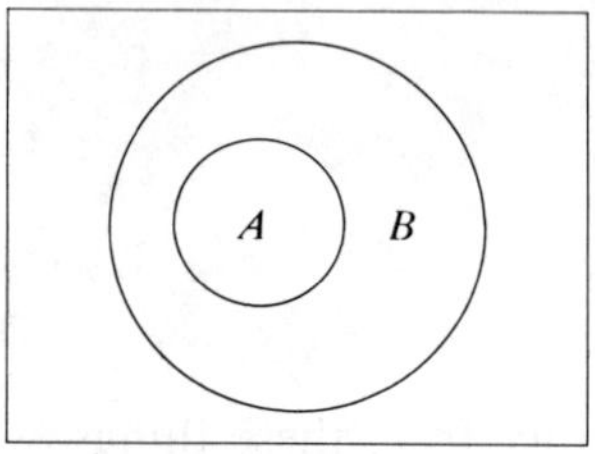

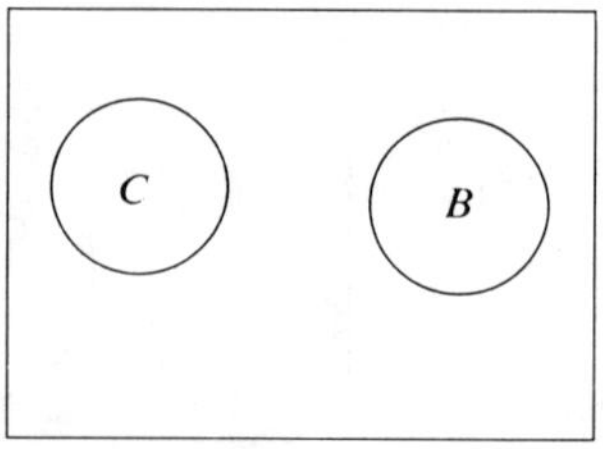

Putting 1 and 2 together, we have:

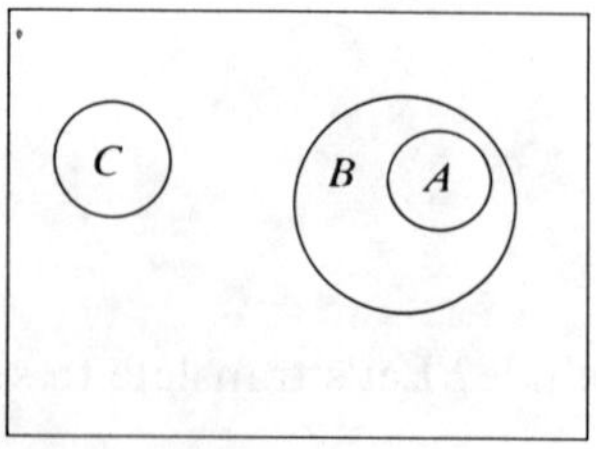

Wow! A is completely away from C so $A \cap C = \emptyset$, and no A is C. Try the following for yourself.

1. Determine the validity of the following so-called arguments, and their conclusions.

a. Given: 1. All A is B
2. Some A is C

Conclusion: Some B is C

b. Given: 1. No A is B
2. Some A is C

Conclusion: No C is B

c. Given: 1. No A is B
2. Some A is C

Conclusion: Some B is C

d. Given: 1. No A is B
2. Some A is C

Conclusion: Some A is not C

e. Suppose 1–3 are given items:

1. Some A is B.
2. Some C is B.
3. No D is B.

Find from the following all correct conclusions, if any.

1. Some C is A.
2. No C is A.
3. Some D is not A.
4. Some D is C.
5. No D is $A \cap C$.

Basic Ideas of Boolean Arithmetic

If you have really tried to become well acquainted with the above ideas, certain things should suggest themselves to you. First notice the similarity between set union and the arithmetic operation of addition. Is there a similarity here with the sentential connective word "or"? Boole and DeMorgan noticed these. They also noted that set intersection behaved as the word "and," and like arithmetic multiplication. Finally, complementation is the analogue of "not" and arithmetic subtraction. However, the rules governing Boolean arithmetic and numeral (ordinary) arithmetic are not quite the same.

Let us look at some examples of Boolean arithmetic. The first major difference between numeral and Boolean arithmetic is that the latter has only two digits $\{0, 1\}$ upon which to build symbol strings, while the former has ten $\{1, 2, 3, \ldots, 9\}$. Thus there are fewer Boolean arithmetic "facts" to learn. The following tables give these facts.

+	0	1
0	0	1
1	1	1

	0	1
0	0	0
1	0	1

−	0	1
0	1	0
1	1	1

Now several things about these tables look strange. First we notice that in Boolean arithmetic $1 + 1 = 1$ and $1 - 1 = 1$. Remember, this is a definition. Why it was defined this way is a different question. It might be best at this point to consider these tables as axioms with $+, \cdot, -, 0$, and 1 as undefined terms. Now, let us look at our "subtraction" table. This one is really strange. However, at the risk of adding to the confusion, let us define $\overline{0}$ to be 1 and $\overline{1}$ to be 0. Now notice

$$0 + \overline{0} = 0 + 1 = 1$$

$$0 + \overline{1} = 0 + 0 = 0$$

$$1 + \overline{0} = 1 + 1 = 1$$

$$1 + \overline{1} = 1 + 0 = 1.$$

These lines can be interpreted as expressing the Boolean subtraction facts with the following definitions:

$$0 + \bar{0} = 0 - 0$$
$$0 + \bar{1} = 0 - 1$$
$$1 + \bar{0} = 1 - 0$$
$$1 + \bar{1} = 1 - 1.$$

Thus suppose we have two Boolean strings $A = 0011$ and $B = 0101$. To add these we use the table and add as though they were ordinary numerals:

$$\begin{aligned} A&:0011 \\ B&:\underline{0101} \\ A + B&:0111. \end{aligned}$$

In a similar manner "multiplication" is done digit by digit:

$$\begin{aligned} A&:0011 \\ B&:0101 \\ A \cdot B&:0001. \end{aligned}$$

Longer strings can be handled in much the same way:

$$\begin{aligned} A&:10101001 \\ B&:\underline{00101110} \\ A + B&:10101111 \\ A \cdot B&:00101000 \end{aligned}$$

Notice that the strings so far have to be of the same length. Try to define or devise a way of dealing with strings of unequal length if it is possible.

Exercises

A. Find Boolean sums and products of the following strings of Boolean digits.

1. A: 0011
   B: 1100
2. $A = 0101$
   $B = 1010$
3. $A = 00110011$
   $B = 01010101$

B. For a string A of Boolean digits define $\bar{A}$ to be the result of interchanging 0's and 1's. That is,

$$\text{if } A = 0011, \text{ then } \bar{A} = 1100;$$
$$\text{if } B = 1001, \text{ then } \bar{B} = 0110.$$

Define I to be the string with all 1's, and 0 the string with all 0's. Show in as much generality as you can that:

1. $A + \bar{A} = I$
2. $A \cdot A = 0$
3. $A + I = I$
4. $A + 0 = A$
5. $A + A = A$
6. $A \cdot I = A$
7. $A \cdot 0 = 0$
8. $A \cdot A = A$
9. Using these facts show that $S = (I, A, \bar{A})$ is not a group under Boolean addition or multiplication. There are, however, some groups to be found if you look only at pairs of strings. For example (I, A) forms a group under both operations. See if you can find others.

C. Let $A = 0011$ and $B = 0101$. Then $A + B = 0111$. Thus, $\overline{A + B} = 1000$. What are

1. $\bar{A} \cdot \bar{B}$?
2. $\overline{A \cdot B}$?
3. $\bar{A} + \bar{B}$?

D. Using the same strings for A and B, verify the following.

1. $A \cdot B + A \cdot \bar{B} = A$
2. $A + A \cdot B = A$
3. $(A + B) \cdot (\bar{A} + \bar{B}) = A \cdot \bar{B} + \bar{A} \cdot B$

E. Let $A = 00001111$, $B = 00110011$, and $C = 01010101$. Show that the following relations hold.

1. $A \cdot (B + C) = A \cdot B + A \cdot C$
2. $A + B \cdot C = (A + B) \cdot (A + C)$

Back to Sets

As before we wish to denote a universal set by a rectangular region. We wish to slightly change our picture of a subset however. This time we shall view subsets as being cells of the rectangle. Let A and B be sets in some universal set $\mathscr{E}$.

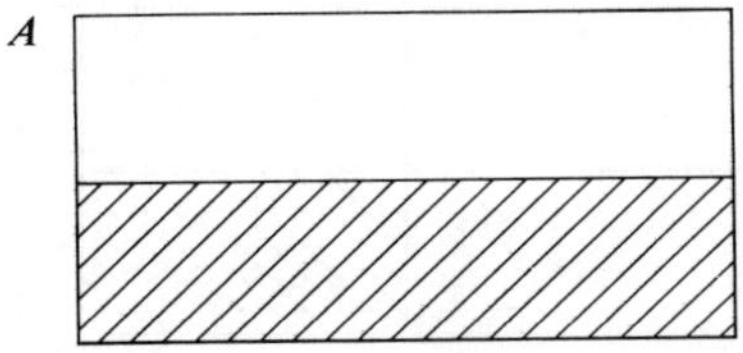

$\bar{A}$

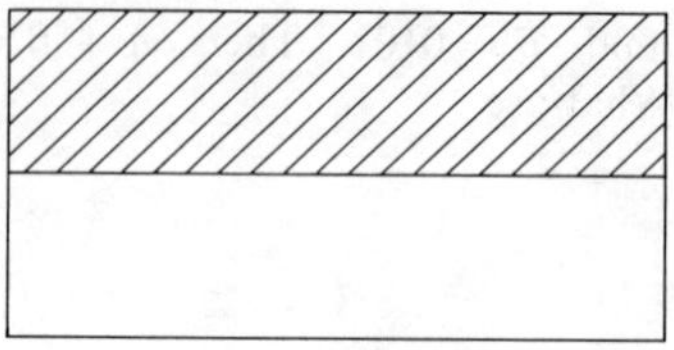

B

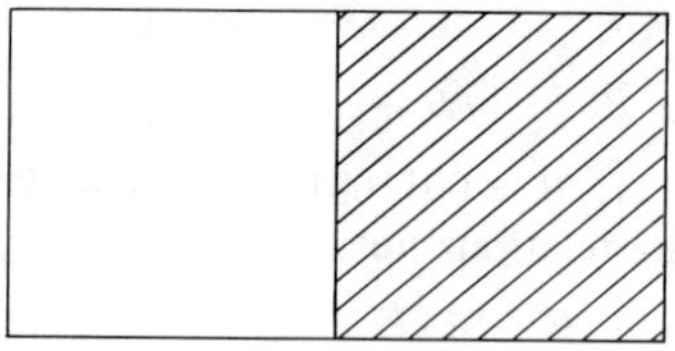

$\bar{B}$

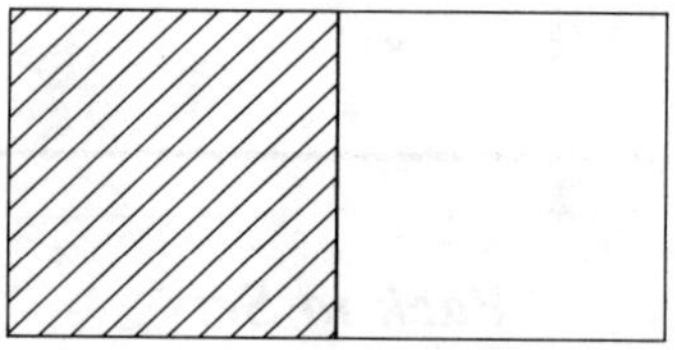

The diagrams say that the bottom half will always denote A, while the top half will denote not-A. The right-hand portion will denote B, while the left-hand portion will denote not-B. Thus, $A \cap B$ is given by

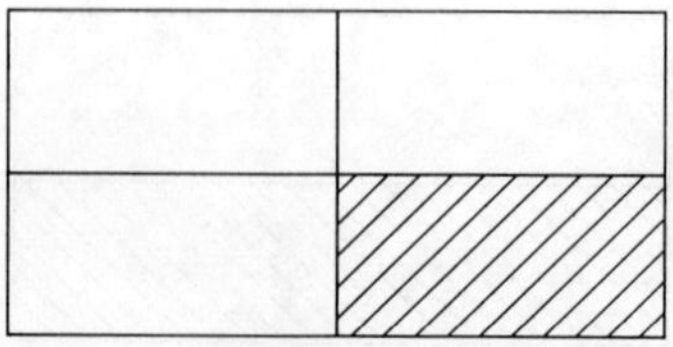

On the other hand, $A \cup B$ becomes

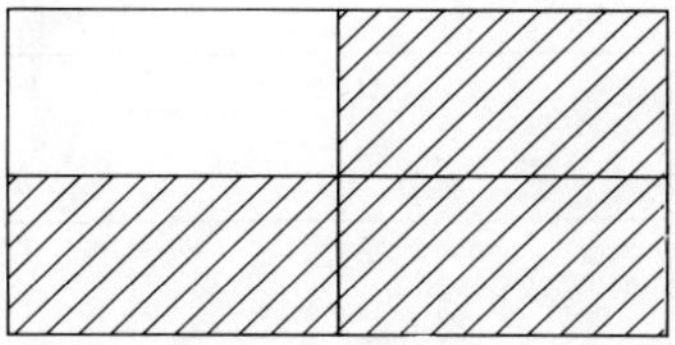

Rather than shading it, it is perhaps easier merely to label the areas:

$\bar{A} \cap \bar{B}$	$\bar{A} \cap B$
$A \cap \bar{B}$	$A \cap B$

As an exercise indicate by shading the following regions:

1. $\bar{A} \cup \bar{B}$
2. $(A \cap \bar{B}) \cup (\bar{A} \cap B)$
3. $(A \cap B) \cup (\bar{A} \cap \bar{B})$
4. $(A \cup B) \cap (\bar{A} \cup \bar{B})$
5. $(A \cup \bar{B}) \cap (\bar{A} \cup B)$

Make a note of any equivalent regions.

For two sets A and B we see we must have four cells. For convenience let us number the cells as follows:

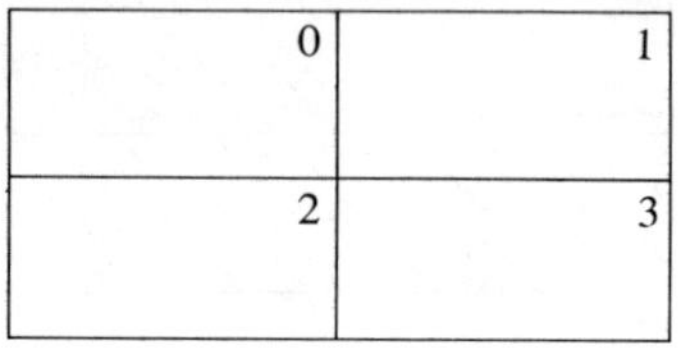

Refer to the diagram for A and notice that cells 2 and 3 are shaded whereas cells 1 and 3 are shaded for B. Let us arrange these cells in a row.

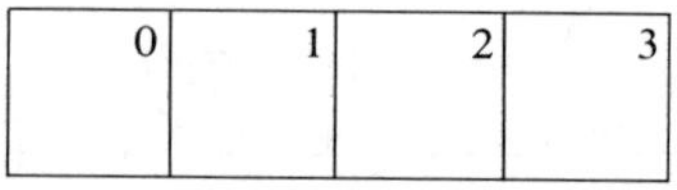

Then A looks like:

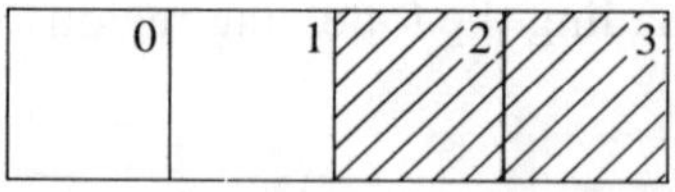

This means there is nothing in cells 0 and 1 and there is something in cells 2 and 3. If we indicate nothing by 0 and something by 1, then A can be written as

$$A = 0011,$$

and

$$B = 0101.$$

Look at the previous section! It doesn't take much work to show that $A \cap B$ can be represented by the string of symbols 0001, while $A \cup B$ becomes 0111. This, of course, means that $A \cup B$ can be interpreted as $A + B$, and $A \cap B$ can be interpreted as $A \cdot B$. Hence, Boolean arithmetic or algebra applies to sets and thus to logic as we presented it above. We shall also give some further examples in later sections.

Three or More

The situation becomes somewhat more complex when there are more than two items. For one thing, with the addition of a single set, say C, we must divide the

A

0 0	1 0	2 0	3 0
4 1	5 1	6 1	7 1

B

0 0	1 0	2 1	3 1
4 0	5 0	6 1	7 1

C

0 0	1 1	2 0	3 1
4 0	5 1	6 0	7 1

universe into eight cells. This is, of course, double the number needed for two sets. Indeed, for four sets we would need 16 cells, and for five sets, 32 cells. In general for n sets we would need 2^n cells. That gets messy. However, let us see a possible arrangement for three and four sets. We will leave the case of five or more to those readers who delight in such things.

The numbers in the upper right-hand corner of each cell denote the cell number. The digit 0 or 1 in the lower left-hand corner indicates nothing or something respectively as before. Notice now what happens if we combine all three diagrams into one single diagram:

0 000	1 001	2 010	3 011
4 100	5 101	6 110	7 111

How does one interpret the lower left-hand string of digits? For example, let us look at 110 in cell 6. This says that we have some A and some B but no C. In set language cell 6 represents $A \cap B \cap \bar{C}$, or in Boolean notation $A \cdot B \cdot \bar{C}$. It is interesting to observe that 110 is the binary representation for the decimal numeral 6. (Is this an accident?)

Let us now use this to check the validity of an argument.

Given: 1. All A is B.
2. All B is C.
Conclusion: All A is C.

By using a diagram, we can test to see whether the proposed conclusion can be obtained from the given items though it does not seem as straightforward as the Euler-Venn diagrams.

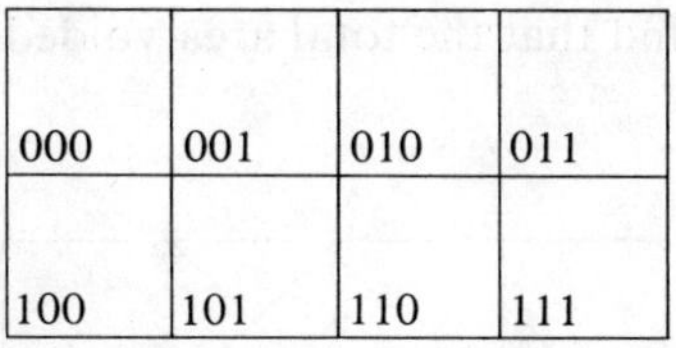

We proceed by voiding the cells which are impossible. Thus, we have

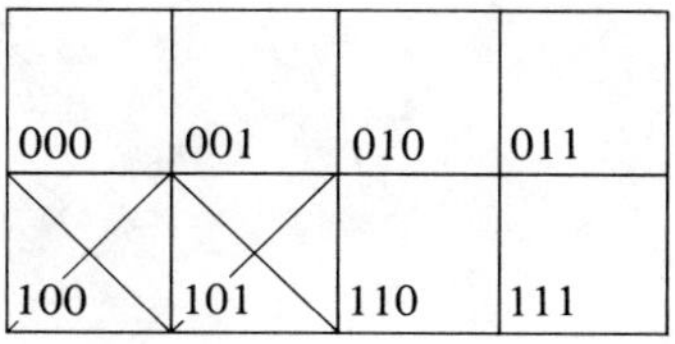

for the first item.

Since we are also given that "All *B* is *C*," the second diagram is given by

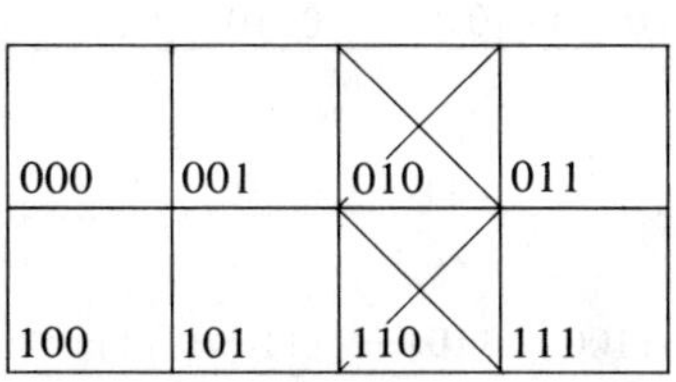

Putting these together we find that the total area voided is

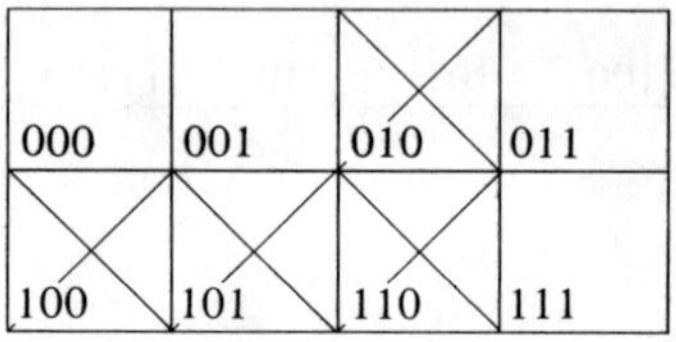

Now the only cell in which an A occurs is one in which both A and C occur. In other words every occurrence of A is one of C. Hence every member of A is a member of C. This says that A is a subset of C; hence, "All A is B" is a correct deduction.

Let us consider a more complex problem with the following sentences given:

1. All T is B;
2. Some T is D;
3. Some F is T.

Our diagram now has 16 cells to denote the sets T, B, D, and F in that order:

0000	0001	0010	0011
0100	0101	0110	0111
1000	1001	1010	1011
1100	1101	1110	1111

Since "All T is B," we can strike out cells 1000, 1001, 1010, and 1011. That much is easy. The real test of understanding comes in dealing with the two items involving the word "some." In the first case where "Some T is D," the cells in which this can happen are 1010, 1011, 1110, and 1111. The first two of these have been eliminated, leaving only 1110 and 1111 as possible. Looking at "Some F is T," we see that cells 1001, 1011, 1101, and 1111 are those exhibiting this characteristic. Of these only 1101 and 1111 are possible. What kind of deductions can be made? Let us list possible cells:

1100,
1101,
1110,
1111.

We cannot use any of the other cells since they are not explicitly mentioned. Interpreting these strings of digits we get:

1. All T is B, which we already know;
2. Some T is F, or some F is T which are equivalent (why?), and already known;
3. Some B is F, or some F is B.

The third statement is something new! Explain in detail why we cannot conclude either "All B is F" or "All F is B."

For further exercises use the ones in the previous section. Try the example by Lewis Carroll given in Pfeiffer's article in "Symbolic Logic" to be found in the readings.

Application of Boolean Theory to Chapter Four

While the above material can be read independently of Chapter 4, the present section is best understood in relation to that chapter. In fact, if you have read Chapter 4 carefully, most of this material you could write yourself. Consequently, we shall present you only with a program for careful analysis.

The propositional forms introduced in Chapter 4 carry with them the tacit assumption of referring to something. That something referred to is usually a set. We have already seen that set intersection behaves like the word "and." Set union behaves like the word "or." Set complements behave like the word "not." Thus, it remains for us to find suitable meaning for "only if." However, before we do this, let us interpret the T's and F's of the truth tables. Let A and B be sets and let p and q be propositional forms specifying A and B respectively. The following table gives us a dictionary for translation:

A	B	$A \cup B$	$A + B$	p	q	$p \vee q$
0	0	0	0	F	F	F
0	1	1	1	F	T	T
1	0	1	1	T	F	T
1	1	1	1	T	T	T

Thus, set union can be translated as Boolean sum, and each can be translated as "or." Here we identify F with 0 and T with 1. As an exercise, make a similar table relating intersection, Boolean product and "and."

To show a correspondence between "only if" and some Boolean binary operation is trivial; for this can be done simply by suitable definition. (Before you read further, try to do just that.) The interesting problem is to try to find some set-theoretic meaning for implication.

We first note that the translation from propositional forms to Boolean forms is accomplished by:

p	q	$p \to q$	A	B	$A \Rightarrow B$
F	F	T	0	0	1
F	T	T	0	1	1
T	F	F	1	0	0
T	T	T	1	1	1

"⇒" will be used as the Boolean correspondent of "→", i.e., "only if." It is helpful to note the following relationships which can be easily (as an exercise) verified:

$$\bar{A} + B = \overline{A \cdot \bar{B}} = A \Rightarrow B.$$

Let's see what our cell table tells us concerning this phenomenon:

00	01
10	11

$\bar{A}$ is the region given by cells 00 and 01 while B is given by cells 01 and 11. Hence, 10 is specifically excluded. This says that A never occurs without also being a B. Thus, A is a subset of B. It is now obvious that the appropriate translation of $p \rightarrow q$ is $A \subset B$ where A is the set specified by p and B is the set specified by q.

Summary

The principal theme in the material just presented, as in everything we have done, is to demonstrate mathematical method in action. Remember in this context the object that we are dealing with is not nearly so important as the method employed. Structure and its analysis are of primary concern.

The rules used in the analysis, the things analyzed, and the results obtained are far from new. The analysis of logical processes by using cells goes back to C. L. Dodgson who is better known as Lewis Carroll. In 1887, he published a book called

"Symbolic Logic, Part I: Elementary," published by Macmillan and Co., Ltd., London. In this country in recent years the book has been reissued by Dover Publications Inc., 1958. This material is highly readable and entertaining, and, of course, educational.

Cell diagrams in preference to Euler-Venn diagrams are finding their way into modern texts on computer science. See the following references:

1. Schneid, Francis. "Introduction to Computer Science," Schaum's Outline Series. McGraw-Hill Book Company, 1970.
2. Pfeiffer, Paul E. *Sets, Events, and Switching*. McGraw-Hill Book Company, 1964.

Index